EARTH WALLS

ELLIS HORWOOD SERIES IN CIVIL ENGINEERING

Series Editors

Structures: Professor H.R. EVANS, Department of Civil Engineering, University College, Cardiff

Hydraulic Engineering and Hydrology: Dr R.H.J. SELLIN, Department of Civil Engineering, University of Bristol

Geotechnics: Professor D. MUIR WOOD, Cormack Professor of Civil Engineering, University of Glasgow

Addis, W.	STRUCTURAL ENGINEERING: The Nature of Theory and Design
Allwood, R.	TECHNIQUES AND APPLICATIONS OF EXPERT SYSTEMS IN THE CONSTRUCTION INDUSTRY
Bhatt, P.	PROGRAMMING THE MATRIX ANALYSIS OF SKELETAL STRUCTURES
Blockley, D.I.	THE NATURE OF STRUCTURAL DESIGN AND SAFETY
Britto, A. M. & Gunn, M.J.	CRITICAL STATE SOIL MECHANICS VIA FINITE ELEMENTS
Bljuger, F.	DESIGN OF PRECAST CONCRETE STRUCTURES
Carmichael, D.G.	STRUCTURAL MODELLING AND OPTIMIZATION
Carmichael, D.G.	CONSTRUCTION ENGINEERING NETWORKS
Carmichael, D.G.	ENGINEERING QUEUES IN CONSTRUCTION AND MINING
Cyras, A.A.	MATHEMATICAL MODELS FOR THE ANALYSIS AND OPTIMIZATION OF ELASTOPLASTIC SYSTEMS
Edwards, A.D. & Baker, G.	PRESTRESSED CONCRETE
Fox, J.	TRANSIENT FLOW IN PIPES, OPEN CHANNELS AND SEWERS
Francis, A.J.	INTRODUCING STRUCTURES: Civil & Structural Engineering, Building & Architecture
Graves-Smith, T.R.	LINEAR ANALYSIS OF FRAMEWORKS
Gronow, J.R., Schofield, A.N. & Jain, R.K.	LAND DISPOSAL OF HAZARDOUS WASTE
Hendry, A.W., Sinha, B.A. & Davies, S.R.	LOADBEARING BRICKWORK DESIGN, 2nd Edition
Heyman, J.	THE MASONRY ARCH
Holmes, M. & Martin, L.H.	ANALYSIS AND DESIGN OF STRUCTURAL CONNECTIONS: Reinforced Concrete and Steel
Hudson, J.A.	ROCK ENGINEERING SYSTEMS: Theory & Practice
Iyengar, N.G.R.	STRUCTURAL STABILITY OF COLUMS AND PLATES
James, A.N.	SOLUBLE MATERIALS IN CIVIL ENGINEERING
Jardine, R. & Hight D.W.	SOFT CLAY FOUNDATION ENGINEERING
Jewell, R.A., Houlsby G.T., Milligan G.W.E.	REINFORCED SOIL
Jordaan, I.J.	PROBABILITY FOR ENGINEERING DECISIONS: A Personalistic Approach
Kutmanova, I., Skaloud, M., Tochacek, M.	PLASTIC DESIGN IN WEBS, PLATES AND SHELLS
Kwiecinski, M.	PLASTIC DESIGN OF REINFORCED SLAB-BEAM STRUCTURES
Lencastre, A.	HANDBOOK OF HYDRAULIC ENGINEERING
Leroueil, S., Magnan, J.P. & Tavenas, F.	EMBANKMENTS ON SOFT CLAYS
MacLeod, I.	ANALYTICAL MODELLING OF STRUCTURAL SYSTEMS
Megaw, T.M. & Bartlett, J.	TUNNELS: Planning, Design, Construction
Megaw, T.M. & Bartlett, J.	TUNNELS: Planning, Design, Construction
Megaw, T.M. & Bartlett, J.	TUNNELS: Planning, Design, Construction, Volume 2
Melchers, R.E.	STRUCTURAL RELIABILITY ANALYSIS AND PREDICTION
Mrazik, A., Skaloud, M. & Tochacek, M.	PLASTIC DESIGN OF STEEL STRUCTURES
Pálossy, L., Scharle, P., Szalatkay I.	EARTH WALLS
Pavlovic, M.	APPLIED STRUCTURAL CONTINUUM MECHANICS: Elasticity
Skaloud, M.	STEEL STRUCTURE STABILITY: Design & Applications of Webs & Flanges
Spillers, W.R.	INTRODUCTION TO STRUCTURES
Vilnay, O.	CABLE NETS AND TENSEGRIC SHELLS: Analysis and Design Applications
Williams, K.	CIVIL ENGINEERING CONTRACTS VOLUME 1: The Law, Administration & Safety
Williams, K.	CIVIL ENGINEERING CONTRACTS VOLUME 2: Planning, Estimating & Tendering
Zloković, G.	GROUP THEORY AND G-VECTOR SPACES IN STRUCTURAL ANALYSIS: Vibration, Stability and Statics

EARTH WALLS

by

LÁSZLÓ PÁLOSSY

PÉTER SCHARLE

ISTVÁN SZALATKAY

all of the Hungarian Institute for Building Sciences

Translation Editor
Dr. T. G. DAVIES
Department of Civil Engineering, University of Glasgow

ELLIS HORWOOD
New York · London · Toronto · Sydney · Tokyo · Singapore

English edition first published in 1993
in coedition between
ELLIS HORWOOD LIMITED
Market Cross House, Cooper Street Chichester, West Sussex, PO19 1EB
England

A division of
Simon and Schuster International Group
and
AKADÉMIAI KIADÓ
Budapest, Hungary

British Library Cataloguing in Publication Data

A catalogue record for this book is available from the British Library

ISBN 0-13-223876-4

Library of Congress-in-Publication Data

Available from the publishers.

This book is the revised English version of the Hungarian *Földtámfalak* published by Műszaki Könyvkiadó, Budapest

Distributed in Hungary, Albania, Bulgaria, China, Czech Republic, Democratic People's Republic of Korea, Mongolia, Poland, Romania, Slovakia, the successor states of the Soviet Union and Yugoslavia and Vietnam by

AKADÉMIAI KIADÓ
P.O.Box 245, H-1519 Budapest, Hungary

Printed in Hungary by Akadémiai Nyomda, Budapest

CONTENTS

1

MECHANICAL BEHAVIOUR OF EARTH STRUCTURES

1.1 SOIL AS A CONSTRUCTION MATERIAL

Building materials emanate from two sources; firstly, natural building materials from the environment, which can be used without alteration or can be improved by a little effort for this purpose and, secondly, building materials which have been manufactured to satisfy constructional, aesthetic and other demands.

The applications of these building materials has motivated research into their physical, chemical and mechanical properties from which better and more economic structures and advanced techniques have emerged.

Besides technical matters, economics (including construction costs) is important. This is why some building materials which are far from ideal for structural purposes have not disappeared from practice even today. Among them we find soil, which in its original condition is often unfit for application and whose treatment is expensive and causes delay. However, soil is widely used for construction purposes and the aim of this book is to introduce the reader to a new field of engineering techniques where soil is the predominant constituent.

The term soil has different meanings in agriculture, mining and in the construction industry. The physical condition and mechanical behaviour of a soil varies widely. When used as a building material, two sorts of soil or two conditions of the same soil may display pronounced differences and the reader is advised to keep this in mind.

1.1.1 Historical background

Soil has been used as a building material since before history. Methods of strengthening earth materials and improving their qualities for use in practice have been used for several thousands of years (*Koerner, Welsh 1980*). For example, from the Holy Bible:

> 'That same day the king of Egypt commanded the slave-drivers and the foremen: "Stop giving that people straw for making bricks. Make them go and find it for themselves"... ' Ex.5(6–10).

Observation of nature have precipitated thoughts on how the strength of soil can be improved by inserting inclusions in natural soil. *Vidal (1966)* was inspired by conifers mixed in sand at a seashore to invent the principle of terre armée (reinforced soil). Natives in the Tropics use bamboo stems to make their tracks passable, and so on.

Study of structural performance, availability of advanced industrial products, improvement in building technology, have all contributed to widen the field for the application of such innovations. One of the successful achievements in recent years was the reinforced earth wall (*Kézdi, 1966; Scharle, Szalatkay, 1977*). Soil nailing and application of root-like piles for foundations (*Dulcz, 1981; Gérusz, 1981; Bányai, 1982*) have also gained favour in industry. All of these ideas and their further modifications and innovations have spawned 'hundred and one' sorts of soil reinforcing methods (*Bartos, 1979*), Fig. 1–1.

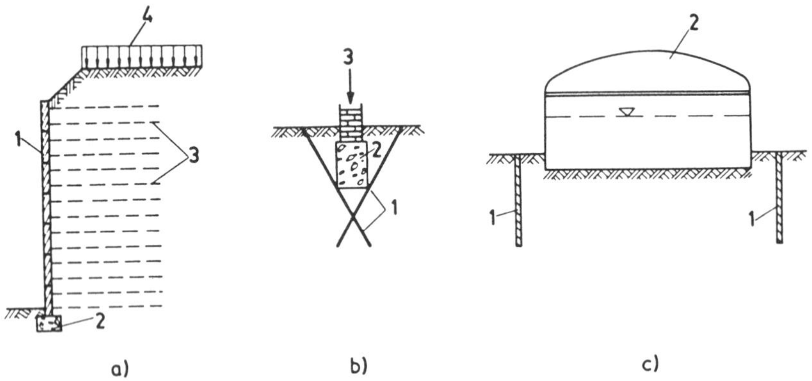

Figure 1–1 Soil strengthening methods

a) reinforced earth wall: *1* face elements; *2* base; *3* reinforcing inclusions; *4* overburden

b) application of root-like piles in the subsoil: *1* root-like piles; *2* footing; *3* pillar, column, or wall

c) obstructing sideward displacement: *1* membrane wall or sheet piles; *2* reservoir

At this point we have to turn to classification and terminology in the context of earth retaining structures. It is not easy to find a common definition for earth-made structures or to give a unique description of them.

Two basic approaches influence this categorisation:

— Resultant loading forces in buildings are generally transmitted into the soil through foundations (footings, piers, piles, etc.). The magnitudes and directions of force transmission are mostly determined by external loading factors; consequently, the mechanical model for calculation can be derived from the theory of structures, with regard to the embedding soil.
— Where inclusions are inserted in the soil mass to modify its deformation conditions and to induce more strength in the mass—independently from acting external forces—, the continuum model pertaining to an inhomogeneous, anisotropic mass is needed for calculation.

These two extreme situations are somewhat simplified, and do not reflect precisely some conditions met in practice. For example, in the case of an anchored retaining earth wall where the dominant force distribution is governed by an anchorage, the rods also modify the stress distribution in the soil mass. A reinforced soil retaining wall is commonly idealized as a gravity earth structure but the face elements exert concentrated tensile forces on the reinforcing strips, which are transferred to the soil mass. To overcome these difficulties, calculations must reflect the dominant influencing factor.

Another factor that may help in the classification of earth-made structures is the distribution of the two main component materials, soil and reinforcement. On the one hand, the distribution is deliberately designed to enhance anisotropy in the earth-made structure, in leaving compressive and shearing stresses to be borne by the soil and tension by the reinforcement. Seldom will we find earth-made structures in which elements are incorporated to take up bending moments. On the other hand, it may be the aim of the design to disperse reinforcing material in a rather dense and uniform manner in the soil, whereby the composite mass becomes thoroughly woven through by it, because in this case adhesion and friction are desired to act among the components. Also the literature is somewhat confused in its terminology in the use of names such as 'earth construction', 'earth structure', 'soil structure', etc. The term soil structure for example has been used for decades as a soil characteristic to describe the texture of the soil and the correlation between soil particles. We shall use the term soil-made structure, reinforced soil, or soil reinforcement, in the hope that the reader will easily distinguish among them in the sense of this book.

Turning back to the scope of this book, we begin with the simplest earth-made structure, that is, the slope whether it be an embankment from hauled material or a cut. In either case, there is a correlation between slope height, inclination and stability. People of ancient times accumulated experience by observing events in nature and derived their rules of thumb for constructing slopes (burial hills, fortifications, channels, mining pits, etc.).

Features in the ground surface, local soil conditions, limited space, aesthetics, etc. may have compelled ancient people to build stable surfaces which were stronger and perhaps steeper than natural slopes but still resistant to various loads, forces, climatic conditions and human interference. It was again experience and practice that led to use stone for constructing walls and retaining walls of high quality. Stone was, however, not available everywhere. Consequently, pebbles, wood, reeds and various plant fibres were mixed into soils to strengthen them where steep walls were desired instead of space demanding slopes. Durable and aesthetic constructions dated to ancient times are prominent examples of this artisanship.

Increased demand for reshaping the ground surface artificially has shed light to shortcomings of some solutions. Among them, it has to be mentioned that:

— a slope (when it is stable) occupies precious land superfluously;
— on cultivated and arable terrain, however, a step-like reshaping of the ground might be preferred to promote production process;
— the construction of a gravity retaining wall is demanding in labour and material, including haulage.

The invention of concrete and reinforced concrete at the beginning of this century was a big step forward in overcoming the setbacks in formerly known wall construction. Later the widespread use of prefabricated elements spurred the development of earth-made retaining structures. Some of them remind us of hydraulic structures (abutments, bank protection, etc.), others are purposefully designed for land protection and gardening. The most recent developments include retained earth masses in the load bearing structures (cantilever walls), or the formation of load bearing frames by the use of load bearing prefabricated elements and infilled soil cores (gravity wall).

In Hungary and abroad, flat land is getting sparse. New residential districts are driven back to the hills. Establishment of green recreation areas, parking lots, etc. also require reshaping of the ground surface. Horticulture demands efficient use of all available land. Mechanical cultivation and land use in agricultural areas, erosion control, surface drainage also make similar demands, not to speak of traffic highways, protection of waterways, etc. All of these activities make use of retaining structures.

The aim of this book is to inform the reader about the most important issues involved and to give a general view on customary and recently developed structures in this field.

1.1.2 Geotechnical principles

Mechanical, structural and even biological interactions between soil and reinforcing materials are induced by the physical, chemical, mechanical and biological properties of the constituent materials. Arrangement and grain distribution of soil particles on one hand and structural characteristics of the reinforcement on the other hand generate new properties which cannot be obtained by simple 'summation' or superposition. A soil cylinder reinforced horizontally by a geogrid may be considered as a typical example: we cannot predict the axial or lateral strains and the unconfined strength of the composite sample from the strength parameters of the soil and the geogrid. If we wish to know the characteristics of the composite sample, we have to test it as a whole. It is not the purpose of this book to study the properties of each component material or to analyse their influence on composite samples. This problem requires study in the field of mechanics of composite materials. This topic is still a fruitful subject for research since it is largely unexplored but may yield potentially promising results. Here, we are limited to the fundamental correlations and principles which govern the behaviour of earth walls and reinforced earth walls.

The basic constituent material of a reinforced earth wall—whence comes its name—is the soil in the civil engineering or agrotechnics sense of the word. That is why it is very important to distinguish between the types and physical properties of soils. The physical properties, the states of soils and the effect of water and porosity will be dealt with in what follows.

1.1.2.1 Types of soils

From the civil engineering point of view a soil is defined in the Hungarian Standard (MSz) 14.045/4–69 as 'a portion of the earth's crust which is fragmented into small particles of rocks or which has originated from the deposition of volcanic ash'.

Soils can be classified in several ways. Definition of soil types is arbitrary in many cases and somewhat idealized because a soil never fits exactly into one category. Soils can be classified—among other attributes—according to the effort needed to excavate them by hand or machine, according to their compactibility, their geologic origin, or according to their condition or strength. All classifying methods imply in one or another way the particle size distribution of a soil. Conventionally, two major groups are distinguished: coarse-grained and fine-grained soils. Coarse-grained soils are those consisting of particles predominantly larger than 0.02 mm grain size (d) and can be 'poured' when dry. Subgroups within this category are as follows (various national standards may be different to some degree):

$d > 200$ mm — boulder
$d = 20$–200 mm — coarse gravel
$d = 2$–20 mm — fine gravel
$d = 0.5$–2 mm — coarse sand
$d = 0.25$–0.5 mm — medium sand
$d = 0.1$–0.25 mm — fine sand
$d = 0.02$–0.1 mm — fine sand with coarse silt.

The character of soils containing a considerable quantity of grains smaller than 0.1 mm in diameter is influenced by the physicochemical interactions between the particles. Such soils are susceptible to plastic deformation and exhibit properties relating to their sensitivity to moisture content. The classification according to plasticity index I_p (which is explained in section 1.2) is as follows:

$I_p < 5$ — silty fine sand
$I_p = 5$–10 — coarse silt
$I_p = 10$–15 — silt
$I_p = 15$–20 — lean clay
$I_p = 20$–30 — medium clay
$I_p > 30$ — heavy clay.

A soil classification system is illustrated in Table 1–1 in which both the dimensions of particles and the plasticity indices of soil are indicated.

Table 1–1 Soil classification of MSz 14045/4–69

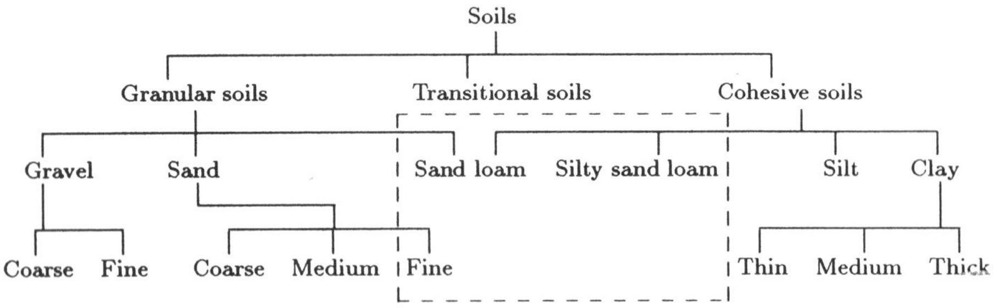

It is evident that most soils cannot be separated along a sharp line into the two main groups. For lack of a better definition, those soils that consist of the soil fractions, fine sand, silty sand and coarse silt are called transitional soils.

The occurrence of these types of soils in Hungary is quite unequal: in three-fifth of the country transitional soils and their geologic varieties (e.g. loess, wind-blown sand) can be found. In the other territories, sands in less proportion and clays in greater proportion occur in Hungary. Most regions of a country and even a construction site is built up geologically of various soil types and their modified versions both on the ground surface and in the stratification. This is why an appropriate investigation on each construction site is unavoidable at a preliminary design stage.

1.1.2.2 Granular soils

Granular soils are generally products of decomposed rocks developing under the influence of weathering. The particles have usually small surface activity. From the shape and size of the particles the origin of the soil can be discerned in many cases. The shape of an individual particle can be angular, laminated, or spherical depending on its exposure to frost, erosion, wearing, crumbling, etc. Although the geotechnic and mechanical behaviour of soils is influenced by all these factors, the most important property for its identification is the distribution of particle sizes. If a soil contains particles not more than 5–6 per cent by weight from the fractions $d < 0.02$ mm (i.e. silt or clay) it is called a granular material. A very small quantity of admixed cohesive soil particles does not influence significantly the mechanical properties of a granular soil, so that this quantity of admixture can be considered as 'contamination'. In the case of greater proportion than this, the presence of the cohesive soil must be taken into account. In such cases description of the soil type is indicated by an adjective (e.g. silty sand).

Granular soils can best be identified by help of the particle size distribution curve. Ordinates of this curve denote the percentage by weight of the fractions finer than the corresponding grain size. A particle size distribution curve is determined in the laboratory by sieve analysis ($d > 0.1$ mm) and a sedimentation (when $d < 0.1$ mm) process. Both methods are discussed in various national standards (E.g. MSz 15045/6–68).

A particle size distribution curve can be described by using only a few data (Fig. 1–2). On the vertical axis the percentage in weight of particles is given which passed the sieve diameter plotted on the horizontal axis in a logarithmic scale. From this curve some important conclusions can be drawn. The uniformity coefficient U is the ratio of grain size at percentage 60% to that at percentage 10%; its value may vary between one and several hundreds. If the diameters of soil particles are nearly equal, then the curve is steep and the uniformity coefficient lies near to unity. Such soils are called 'uniform' or 'poorly graded', and can usually

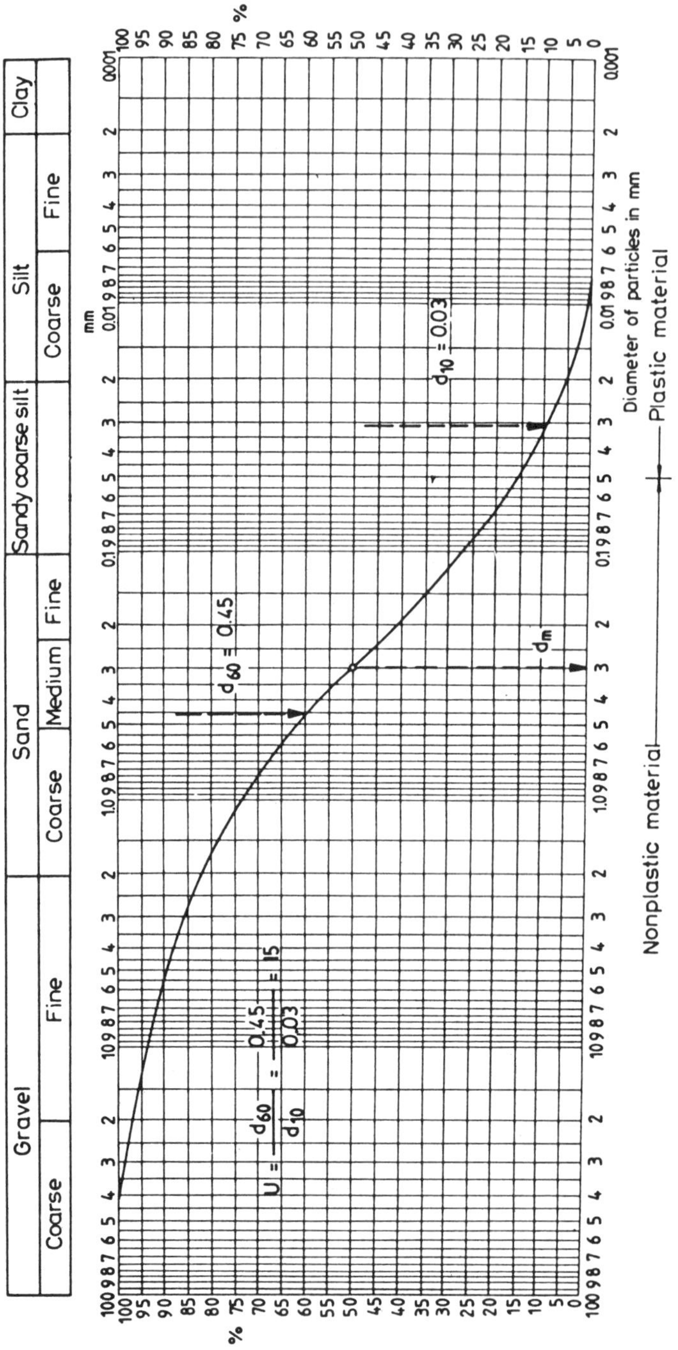

Figure 1–2 Chart for recording particle size distribution

be compacted with difficulty. Their shear strength is low and the soil tends to flow.

As the value of U increases, the curve becomes flatter. These are the so-called well-graded soils. They can be compacted much easier to high densities to achieve greater shear strength and more coherent soil structures.

Another characteristic feature of the curve is its point of inflexion where the diameter d_m can be read. This is the diameter at which the particle size frequency curve attains its maximum. From the particle size distribution curve the composition of a given soil can be illustrated by plotting the triangular diagram shown in Fig. 1–3.

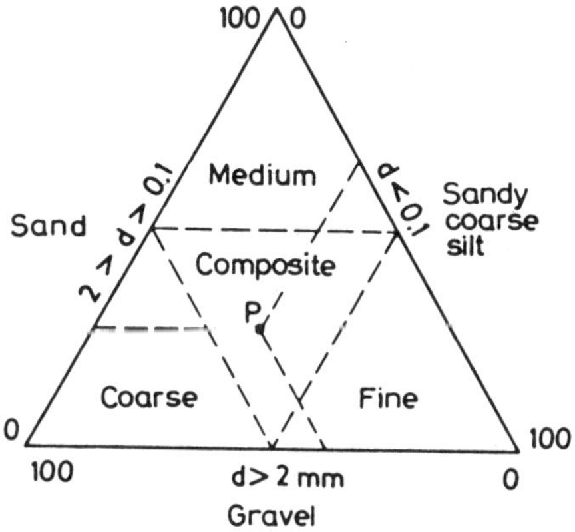

Figure 1–3 Triangle diagram illustrating the phases of a granular mass (P is the point representing a given soil)

1.1.2.3 Cohesive soils

Cohesive soils, to a degree depending on their type, become soft when water is added to them but gain significant tensile and shearing strength in the dry condition. In the dry state they can usually be broken into small pieces only by force, i.e. by cutting, rasping, grinding or sawing. Classification according to particle size gives less information about their mechanical behaviour than the index properties referring to their consistency and sensitivity to water content. In the practice of soil mechanics, three consistency limits (*Atterberg* limits) are accepted:

— liquid limit, w_L,
— plastic limit, w_p,
— shrinkage limit, w_s.

These water contents relate to particular states of consistency which correspond to qualitative changes in soil behaviour. The first two, the liquid and the plastic limit have extraordinary significance in soil mechanics. If a cohesive soil is mixed with water progressively, it becomes a viscous liquid and loses its shear resistance and cohesion. The water content required to reach this state is called the liquid limit (w_L). For exact determination of this parameter, the *Casagrande* apparatus is used (MSz 14045/8–69).

The value of liquid limit ranges between 10...20% for fine sands, 15...30% for silty sands, from 30...40% for silts and 40...150% for clays. These values depend not only on the particle size distribution of the soil but on its mineral composition as well. A cohesive soil with a water content below the liquid limit can deform plastically when subjected to pressure. The water content at which a soil becomes semi-solid is called the plastic limit (w_p). Its value is determined by performing a seemingly arbitrary test whereby the soil is dried by the warmth of the hand in rolling a 3...4 mm thread until it begins to crumble. From the point of civil engineering, optimal conditions for soil compaction occur where the least effort has to be exerted at water content near to the value of plastic limit.

The plastic limit ranges between 17...20% for silty sand, 20...25% for silts and 25...50% for clays.

The difference between the liquid limit and the plastic limit is called the plasticity index ($I_p = w_L - w_p$) which is also an important parameter for identification. Its value can be given by the empirical relationship:

$$I_p = 0.73(w_L - 20) \quad (1\text{–}1)$$

which can be interpreted as the 'A' line in *Casagrande*'s plasticity chart (Fig. 1–4).

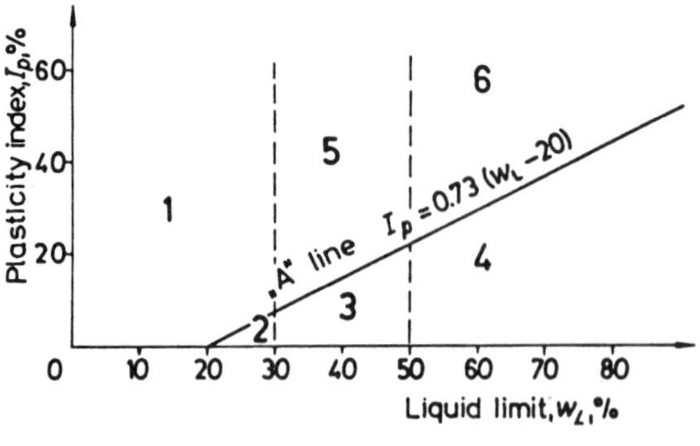

Figure 1–4 The 'A-line' according to *Casagrande*
1 gravel, sand; *2* silty sand, silty or clayey sand;
3, *4* medium and very compressible silts and clays; *5*, *6* medium and highly plastic clays

The chart can be divided into six regions by drawing vertical lines at water contents of 30% and 50% on the abscissa. All regions correspond to certain soil groups or classes and a given soil is named after the region in which the point representing the soil falls (*Kézdi, 1977*).

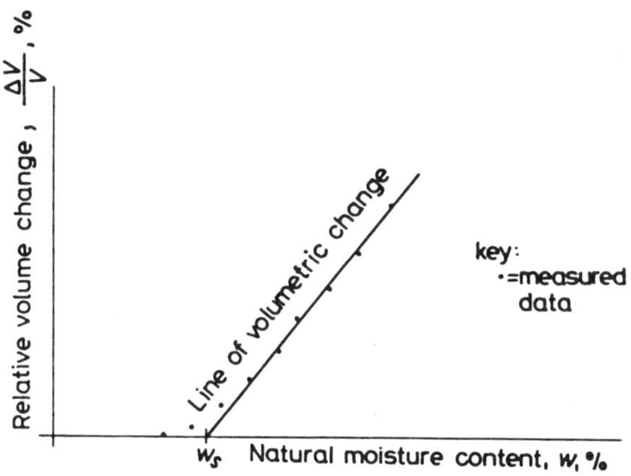

Figure 1–5 Determination of the shrinkage limit by using measured data

If a soil is gradually dried below the plastic limit, a value of water content is reached beyond which no further volume change—i.e. shrinkage—occurs. The water content at this point is called the shrinkage limit (w_s, see Fig. 1–5). A desiccating soil surface shows a characteristic pattern of shrinkage cracks. The finer are the particles of the soil, the greater is the amount of shrinkage. The shrinkage limit is determined by drying out completely a lump of soil and measuring simultaneously its volume and weight (MSz 4045/8–69).

1.1.2.4 Transitional soils

Behaviour of fine sands, silty fine sands and coarse silts (whose average diameter ranges between 0.02 to 0.25 mm, see Table 1–1) is intermediate between granular and cohesive soils. They can be crumbled when dry and moulded when wet and behave as viscous liquids when saturated with water. Classifying them into a separate soil group is reasonable not only because of their transitional character but also for their particular behaviour as a consequence of their physical properties (e.g. compactibility, extreme sensitivity to water content, etc.). These phenomena can be attributed to certain causes, as follows:

Phenomena occurring on the surface of granular particles is related to the presence of free water in the capillaries of soil (based on surface tension) whilst adhesive forces caused by water may be relatively great. No electrochemical bond develops on the surface of granular particles in contradistinction to clays or silts.

— The bonds between soil grains is the greatest at certain water contents and is very sensitive to any variation in water content. Lack or excess of a minute quantity of water compared to this peak breaks the bond between the grains and the soil begins to flow as the bonds lose their strength. The bonding stress in a given soil can be related to the velocity of capillary rise in that soil as shown in Fig. 1–6. where the capillary rise is shown to be a function of average grain size. Capillary head is at a maximum for transitional soils.

— Owing to significant permeability in these transitional soils the pore water pressure dissipates quickly, and so, static consolidation occurs in a short time.

— These soils are highly elastic under dynamic loads, particularly in a critical state where values of saturation attain 85 to 92%. This phenomenon is caused by the elastic deformation of the air bubbles which remain enclosed in the soil due to the moderate value of air permeability which prevents an immediate reaction to quickly alternating dynamic loads. (This phenomenon is very disadvantageous during compaction when the soil behaves like rubber below vibrating rollers and cannot be compacted.)

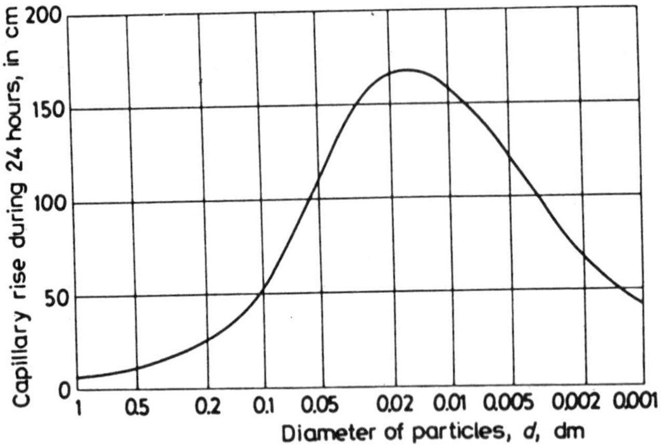

Figure 1–6 Capillary rise during 24 hours versus diameter of particles (*Kézdi, 1979*)

A well-known type of transitional soil is loess. It is a wind blown and sedimented fine sand or silty fine sand which consists of approximately uniform particles (its maximum uniformity coefficient is $U = 10$, with grain sizes between $d = 0.01 \ldots 0.005$ mm). Its liquid limit is

$w_L = 25 \ldots 30\%$, and plasticity index $I_p = 4 \ldots 12\%$. It has an extraordinarily loose structure because the interparticle bonds are formed by a calcareous film and other cementing materials which precipitated when water leached into the holes which were filled with grass and vegetation at the time of deposition. This macroporous structure has some strength in its dry state even under increased overburden pressure but collapses when wetted, owing to the abrupt collapse of the cemented structure. Structural collapse of a loess can easily be investigated in the laboratory by using undisturbed samples. Soaking of soil samples during loading causes a sudden collapse (Fig. 1–7).

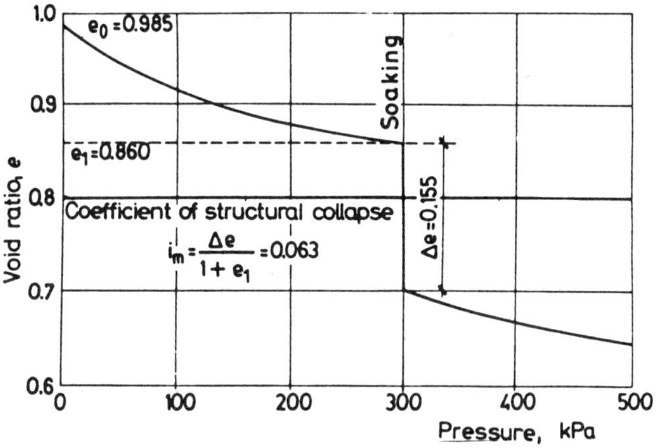

Figure 1–7 Compression curve of a loess; the sample has been exposed to soaking at a load of $q = 300$ kN/m^2

(yellow silty fine sand $w = 6\%$; degree of saturation $S = 0.16$; dry density $\rho = 1.41$ g/cm^3)

1.1.2.5 Composite (mixed) soils

Soils consist generally of composites of fractions of several soil groups of different grain sizes. In such cases, the properties of a given soil can be deduced from those of the composite ones with more or less reliability. In many cases, the reverse of this problem must be solved: from given soil composites of various sources an appropriate soil must be mixed for a given purpose. This purpose can be the improvement of compactibility, reduction in sensitivity, increasing shear strength, etc. The problem of producing artificial soil mixtures is a problem within the field of soil stabilization, especially in mechanical stabilization. For constructing earth roads, the optimum proportions of soil mixtures have to be determined empirically. For instance, if there are two sorts of soils: a gravel with a grain size ranging between 8…16 mm and a sand with a grain size range of 0.5...1 mm, then, to obtain the maximum density and minimum porosity of the desired mixture, proportions of 30 and 70%, respectively, must be taken (Fig. 1–8, *Kézdi, 1967*).

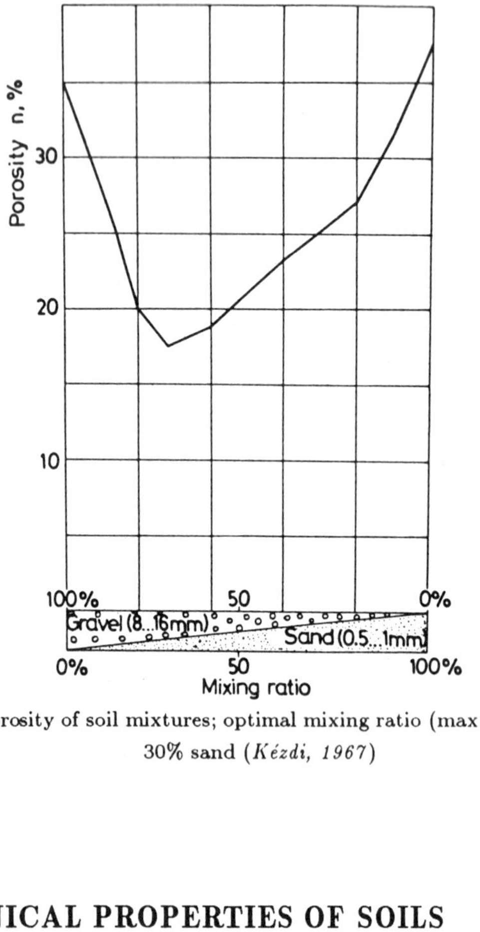

Figure 1–8 Porosity of soil mixtures; optimal mixing ratio (max. density): 70% gravel, 30% sand (*Kézdi, 1967*)

1.2 MECHANICAL PROPERTIES OF SOILS

The mechanical behaviour of soils is influenced by many factors. Durability, strain conditions, dissipation of pore pressures, draining, load bearing and other factors are greatly affected by the states of the soils in addition to their type. The influence of these factors on each other and on soil strength are inexhaustible topics for research in soil physics. Thus, only the methods of investigations and pertinent mechanical parameters will be introduced in the following sections.

To avoid any misunderstanding at the outset it must be emphasized that the phenomenon of soil strength is not connected with the strength of individual particles but refers to the granular mass as a whole and the strain of the individual particles can be ignored. Compressive and shear strain of a soil mass is not the sum of compression and shear strain of individual particles but arises as a consequence of their displacement or rearrangement, i.e. the change of state. The change of state is accompanied by increases or decreases in density and the squeezing out or taking up of water and air. Therefore, the properties of state are discussed as mechanical properties subsequently. In the same way, any change in porosity or moisture content is discussed as a change in mechanical properties as well although we remain aware of the fact which differentiates between the change of state and change in mechanical properties. As any mechanical change is accompanied by a change of state in all cases (however, only the former can be registered) and mechanical properties cannot be analyzed in the terms of state conditions, this approach is intuitively valid.

1.2.1 State properties of soils

Soils are in general three-phase materials consisting of solid, liquid (water) and gaseous (air) phases. In fully saturated or completely dry states, soils are two-phase materials. The proportion of all three phases is determined when only two data are given (since their sum is unity). For example, the specific volume of water and solid material defines the composition of mass. To characterize the state of a granular soil besides the fundamental geotechnical attributes, the relative volume of each composite part must be given (according to the Hungarian Standard MSz 14045/4-69). The relative volumes of the solid (s), water (w) and gaseous (l) phases can be interpreted as shown in Fig. 1-9. The phase composition can be demonstrated using the triangular (phase) diagram (Fig. 1-10).

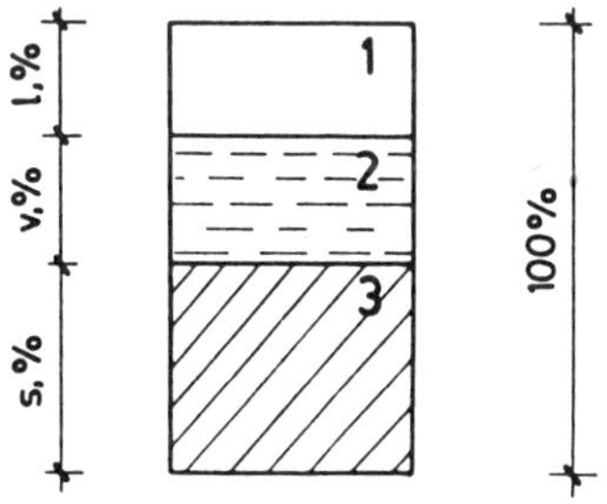

Figure 1-9 Specific volumes of soil phases

1 air; *2* water; *3* solid material

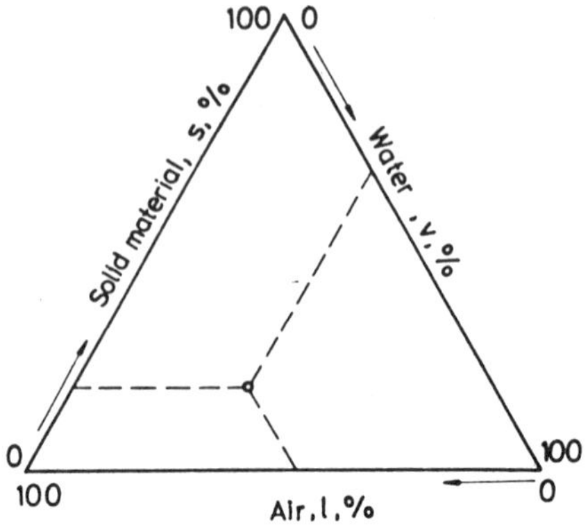

Figure 1–10 Phase composition in triangular diagram

The state of a cohesive soil can be qualified in a simple way by reporting its natural moisture content and its *Atterberg* limits. The consistency index is defined as the ratio of the difference between the liquid limit and the natural water content to the difference between the liquid and plastic limits. In symbols (Table 1–2)

$$I_c = \frac{w_L - w}{w_L - w_P} \tag{1-2}$$

Table 1–2 Soil condition defined with respect to the consistency index, I_c

I_c Index of consistency	Characteristic behaviour
< 0.25	very soft
0.25 – 0.50	soft
0.50 – 0.75	easy to deform
0.75 – 1.00	malleable
1.00 – 1.50	hard
1.50 <	very hard

In practice, by using the quantities determined in the laboratory, many other useful parameters have been derived for characterizing soils. Among them, the most important ones are as follows (see Hungarian Standard MSz 14045/4–69 and 14045/7–69):

— ρ_s specific mass density of solids

— ρ_v specific density of water

— natural moisture content

$$w = \frac{\text{weight of water in the soil}}{\text{weight of soil in dry state}}$$

— void ratio

$$e = \frac{\text{volume of voids}}{\text{volume of solid}}$$

— porosity

$$n = \frac{\text{volume of voids}}{\text{total volume}}$$

— specific mass density of soil ($0 < S < 1$)

$$\rho = \frac{\text{density of soil}}{\text{total volume}}$$

— specific dry mass density of dry soil

$$\rho_\mathrm{d} = \frac{\text{dry mass density of dry soil}}{\text{total volume}}$$

Knowing some basic data (any two of s, v, l, and both ρ_s and ρ_v) all other quantities can be calculated. Many further classifications are based on the values determined by this method, e.g. any soil can be qualified by its grade of saturation (Table 1–3).

Table 1–3 Determination of soil states by means of their grade of saturation (according to the Hungarian Standard MSz 14045/4–69)

Saturation, s	Characteristic
< 0.3	moist
0.3 – 0.8	wet
0.8 – 0.99	close to saturated
1.00	saturated

Density of granular soils can be characterized by other quantities:

— degree of compaction

$$T_{\mathrm{r}\rho} = \frac{\text{specific mass density of dry soil}}{\text{specific maximum mass density of dry soil}} = \frac{\rho_\mathrm{d}}{\rho_{\mathrm{d\,max}}}$$

(the maximum unit weight of a dry soil can be determined by carrying out a *Proctor*-test, according to the Hungarian Standard 14045/11–70),

— relative density

$$T_\mathrm{re} = \frac{e_\mathrm{max} - e}{e_\mathrm{max} - e_\mathrm{min}},$$

where e_{max} is the void ratio of the soil in its loosest state, e_{min} is the void ratio in the densest state obtained in the laboratory, and e is the void ratio in the actual state under examination.

Mechanical behaviour of a soil is affected mainly by its water content and porosity. For coarse granular soils, almost the only determinant property is porosity, because water content has only a slight role. With decreasing grain size, however, the significance of water content increases; in a certain range ($S = 0.7 - 0.87$) a significant increase in strength can be observed in fine granular soils. In such cases, the water adheres to the contact points of the particles and the soil mass becomes bound together by surface tension.

In the case of cohesive (and transitional) soils, the physical situation differs from that described for granular materials. As a function of water content, the consistency of these soils also change. The compressibility and strength of such soils is influenced by consistency index I_c and porosity. The smaller is the grain size of the soil, the greater will be the influence of the consistency index on the strength of the soil; this phenomenon is more obvious in the case of transitional soils. The state (or strength) of a cohesive soil can be estimated from its void ratio and plasticity index using the chart given in (Fig. 1–11).

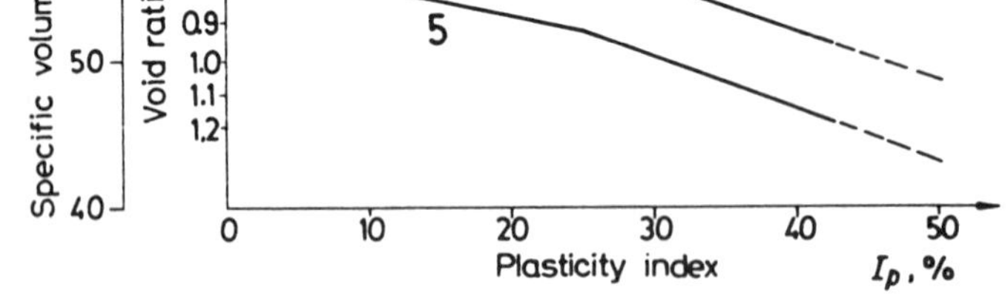

Figure 1–11 Relationship between void ratio, plasticity index and specific volume of solid phase (MSz 14045/4–69)

1 very dense; *2* dense; *3* medium; *4* loose; *5* very loose soil

With respect to compaction, a characteristic relationship exists between the water content and the porosity of soils. For a given compaction effort, different porosities (void ratio) and dry densities can be obtained when the moisture content is varied. The dry density has a maximum value ($\rho_{0,max}$) at a specific water content called the optimum moisture content (w_{opt}). The magnitudes of these two data depend on the way and effort of compaction and are fundamental properties of the given soil when determined by the standard *Proctor* compaction test (according to the Hungarian Standard MSz 14045/11–70), which specifies the compaction effort (including the number and drop height of the blows and the weight of the hammer). The modified *Proctor* test of higher compaction effort corresponds better to field conditions. The relationship between dry density and moisture content for some types of soil is shown in Fig. 1–12.

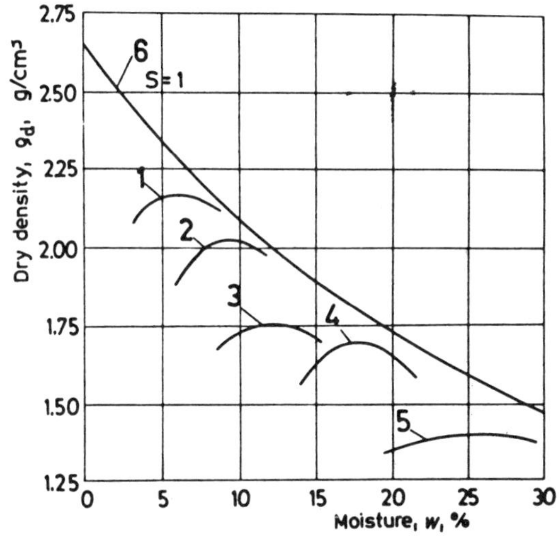

Figure 1–12 Relationship between water content and porosity according to the *Proctor*-test for different types of soil

1 clayey sand; *2* silty sand; *3* sand; *4* silt; *5* fat clay; *6* line of saturation $S = 1$ (*Kézdi, 1976*)

The optimum water content is in the range $w_{opt} = 4 \ldots 12\%$ for granular soils and $w_{opt} = 8 \ldots 20\%$ for cohesive soils. The maximum dry density depends on the mean grain diameter and uniformity coefficient of a soil; e.g. a silty sand (with a uniformity coefficient $U = 5$) has a maximum dry density $\rho_{0,max} = 1.90 \ldots 2.05$ t/m^3, a fine sand has a $\rho_{0,max} = 1.7 \ldots 1.8$ kN/m^3 (with $U = 1.2 \ldots 2.7$).

1.2.2 Strength of soils

One of the fundamental objectives of soil mechanics is the determination of the strengths of soils. Knowledge of strength properties is needed to give proper answers to the following questions:

— allowable loading forces acting on structures,
— stability and bearing capacities of structures embedded in subsoils;
— deformation and displacements of loaded soil masses and structures.

Interpretation of strength and experimental determination of characteristic parameters is a major task. The variety of experimental methods and pertinent limiting conditions are frequently sources of difficulty. Mechanical strength parameters for soils can be determined by applying:

— soil classification methods with respect to state properties of the soil by using diagrams and tables mostly based on empirical knowledge;
— field tests; when results and parameters obtained from load tests and penetrometer or pressuremeter tests are obtained.
— laboratory tests; when stress–strain relationships are measured in specimens taken in the field using laboratory instruments.

In the following, strength concepts connected primarily with laboratory tests are outlined since they are:

— well known and more or less standardized,
— easily reproduced,
— mostly analogous to the elementary state of strength (unconfined or triaxial compression test, shearing and tension tests, etc.),
— suitable to illustrate the relationships between the soil classification methods or state properties and the mechanical parameters.

However, the disadvantages of laboratory tests must be emphasized too. Sampling procedures disturb the initial stress state in the soil and this fundamental effect cannot be eliminated. The initial state of in-situ stress is unknown and, therefore, it can be reproduced only approximately in the laboratory prior to investigation. Due to the small size of the specimens in general, the boundary conditions of a laboratory apparatus differ significantly from that in the field and this may cause substantial (factor of two) differences between measured and real values of strength parameters. In soil mechanics, however, these discrepancies are more or less tolerated, because the inhomogeneity and variability of strength in the soil mass are of the same order.

Based upon the previous discussion, strength properties are discussed in what follows simultaneously with laboratory tests. Some of the most important and simple testing methods are illustrated in Fig. 1–13.

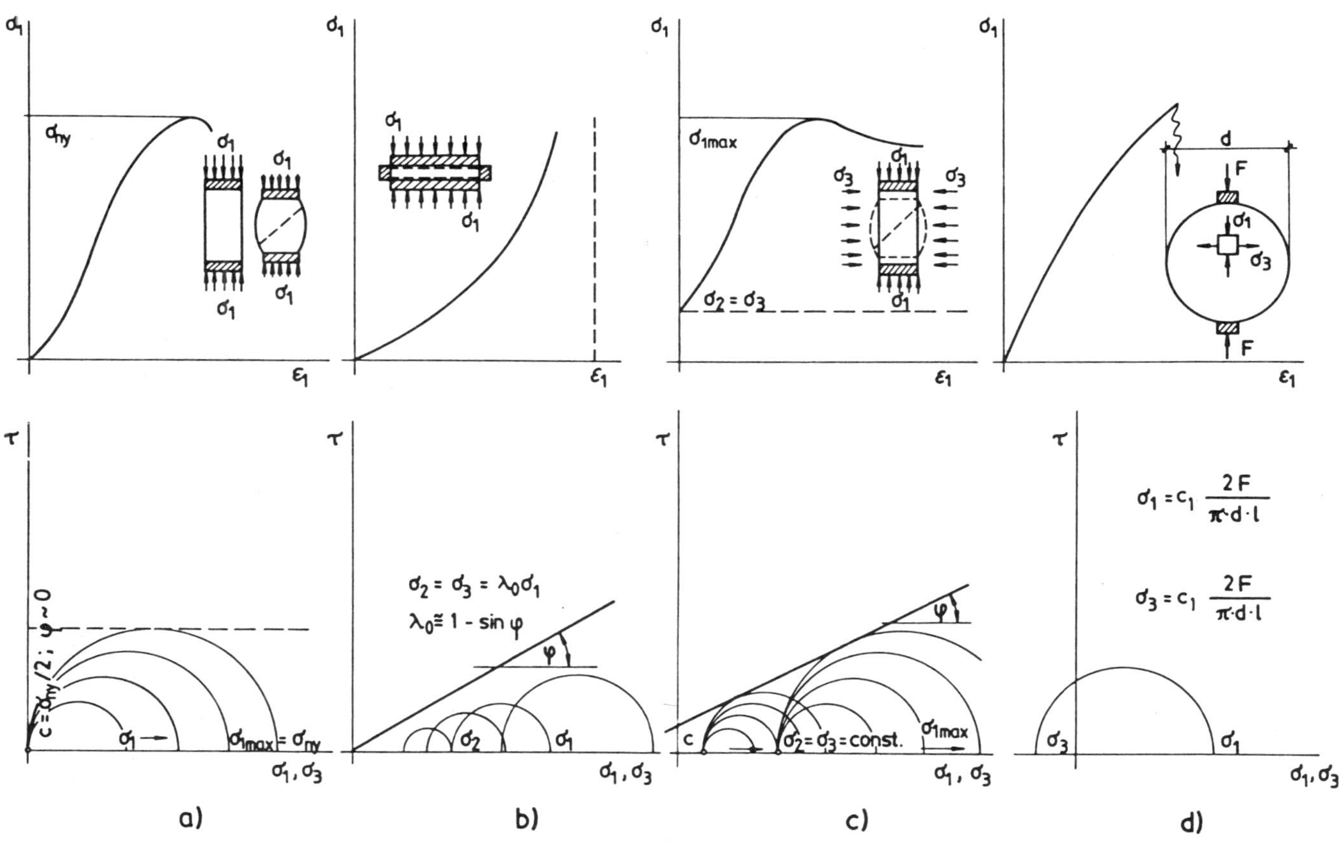

Figure 1–13 Simple laboratory testing methods of soils

a) unconfied compression test; *b)* confined compression test (in oedometer); *c)* triaxial compression test; *d)* split testing; ε_1 = axial strain; τ = shear stress; $\sigma_1, \sigma_2, \sigma_3$ = normal stresses; σ_1 = principal axial stress; l length of the sample

1.2.2.1 Shear strength

Relative displacement of soil particles in a soil mass is the result of shear stresses and in general this can be connected to a state where shear resistance is exceeded. In fact, a soil sample subjected to compression fails by shearing, whereby failure occurs inside the specimen along sliding surfaces.

In soil mechanics, the simple concept of *Mohr–Coulomb*'s failure mechanism is adopted:

$$\tau = a\sigma + b \tag{1–3}$$

according to which the shear strength on a given surface depends linearly on the normal stress acting on the same surface.

Though the relationship (1–3) simplifies radically the problem of the shear strength condition in a multiphase system, it is a good approach and satisfactory solutions can be obtained in many cases provided that certain limitations and complementary criteria are recognized (*Kézdi, 1977*).

In soil mechanics, the parameters a and b can be interpreted explicitly. Instead of coefficient a a term describing the internal friction angle φ is introduced:

$$a = \tan\varphi$$

and b means the cohesion: $b = c$. These two parameters have great significance in soil mechanics; by knowing these, limit state analysis based on the *Mohr–Coulomb*'s failure theory can be performed.

Internal friction can be thought of as sliding and rolling of the particles and, therefore, as the rearrangement of soil structure. Cohesion, however, is due to electrostatic and capillary interaction between the particles depending greatly on the distance between the particles. Hence, it is easy to see that shear strength is highly influenced by the class of the soil and its state properties. For granular soils, the particle size distribution and form of the particles and the value of porosity are important. In the case of large particles, water plays no role at all, while, as the mean grain size decreases and the proportion of cohesive soil increases, the water content becomes more and more significant. For cohesive soils the porosity is an important factor as well, but because the void ratio is influenced by the moisture content, the strength of clay and silt can be assessed directly by knowing the water content.

Only fine-grained soils comprising silt and clay have cohesion. The thickness of the water film adhering to the particles decreases the molecular (electrostatic) bond, thus the cohesion decreases when a soil becomes wet and increases when it becomes dry. In transitional soils (fine sand and silty fine sand) cohesion between particles, as mentioned before, originates from capillary effects. This type of cohesion, due to surface tension and capillary forces, disappears, however, after saturation, inundation or drying.

An ideal soil type, in respect of shear strength and bearing capacity, can be produced by appropriately mixing cohesive and coarse-grained soils (mechanical soil stabilization). A material like this does not lose its strength as its water content changes, since the internal friction is not influenced by wetting or drying, and strength remains even if there is no normal pressure because its shear strength is the result of cohesion. Various values of the internal friction angle, depending on soil state, are shown in Table 1–4. For granular soils the friction angle varies between 20 ... 40° and water content influences its value only when some silt forms part of the soil. The influence of porosity (dry density) is great. In the intervals of $0.7 < I_c < 1.1$ and $10 < I_p < 50\%$ the friction angle of a cohesive material can be estimated from the formula

$$\varphi = 30° - 0.4\, I_p.$$

Cohesion in a dry, granular material is equal to zero. Only soils with a minimum value of plasticity index $I_p = 10\%$ have real cohesion but the value is, however, influenced by the consistency index. Cohesion can range widely from 10 to 1000 kN/m^2 (Fig. 1–14).

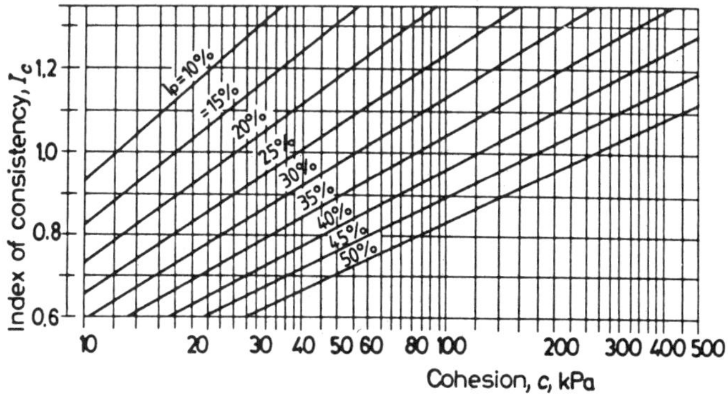

Figure 1–14 Empirical correlation between consistency index I_c, plasticity index I_p, and cohesion c (*Rózsa, 1977*)

Shear strength including friction angle and cohesion can be measured by the direct shear test. The shear box consists of two square or circular metal boxes (Fig. 1–15*a*) placed on each other; one is fixed while the other is moved in horizontal direction. Having loaded the specimen vertically in advance, shearing force is applied through the upper part of the box by pulling it until the sample fails. Horizontal and vertical displacements as well as applied forces are measured. Results may show either densification or loosening of the specimen. Having carried out at

Table 1–4 Typical values of friction angle in soils of various densities (*Rózsa, 1977*)

Soil				Dry or moist		Wet, very wet		Under water	
type	name		ρ_d, g/cm^3	ρ_n, g/cm^3	φ^o	ρ_n, g/cm^3	φ^o	ρ_n, g/cm^3	φ^o
Granular (cohesionless) soils	Coarse gravel fine gravel		1.77 1.93	1.8 2.0	35 40	1.8 2.1	35 40	1.1 1.2	35 40
	Sandy gravel without silt		1.93 2.25	1.9 2.1	32 38	2.0 2.2	32 38	1.2 1.4	32 38
	Coarse and medium sand, with smooth grain size distribution		1.61 2.09	1.7 2.0	32 35	1.8 2.1	32 35	1.0 1.3	32 35
	Silty sand of mixed grain size		1.61 1.93	1.6 2.0	28 32	1.8 2.1	26 30	1.0 1.2	26 30
	Fine silty sand of smooth grain size distribution		1.45 1.93	1.5 1.9	30 34	1.7 2.0	30 34	0.9 1.2	30 34
	Fine sand of mixed grain size (without silt)		1.77 2.09	1.8 2.0	25 28	1.9 2.1	22 25	1.1 1.3	22 25
	well graded sand loam		1.44 1.92	1.5 1.9	25 27	1.7 2.1	22 25	0.9 1.2	22 25
	Silty sand loam		1.44 1.92	1.5 1.9	23 26	1.8 2.1	20 23	0.9 1.2	20 23
		I_p, %		$I_c > 1$		$I_c = 0.5\ldots1$		$I_c < 0.5$	
Cohesive soils	Silt	7...15	1.35 1.60 1.80 1.94	1.5 1.7 2.0 2.2	20 23 25 27	1.7 1.9 2.1 2.3	16 19 22 25	1.9 2.1 2.3 —	12 16 19 —
	Thin clay	15...25	1.38 1.62 1.83 1.97	1.6 1.8 2.0 2.2	16 18 20 23	1.8 2.0 2.1 —	10 15 18 —	2.0 2.1 — —	8 13 — —
	Thick clay	> 25	1.40 1.65 1.87 2.00	1.6 1.8 2.1 2.3	14 16 18 20	1.9 2.1 2.2 —	5 9 11 —	2.1 2.2 — —	3 5 — —

least three measurements and plotting the normal stress-shear strength results in a Cartesian diagram, the *Mohr* envelope (*Coulomb*-line) can be

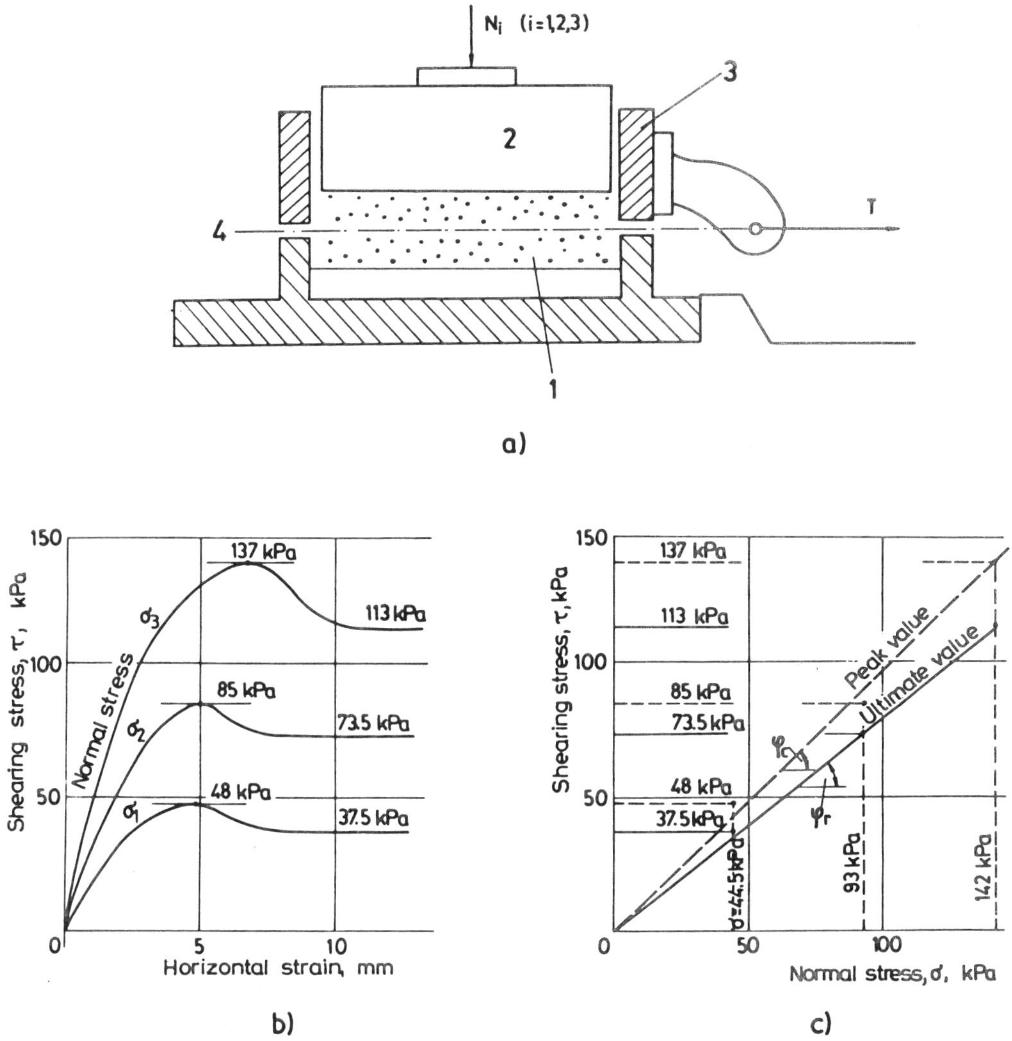

Figure 1–15 Direct shearing test (*Kézdi, 1977*)

a) sketch of the shear box;

b) relationship between shear stress and strain for different normal stresses; *c)* Coulomb's envelope

1 soil; *2* loading plate; *3* moving frame; *4* shear plane;

(coarse sand, $e = 0.59 - 0.62$, $\varphi_{max} = 44°$, $\varphi_r = 35°$, T = pulling force)

constructed. The internal friction angle is the inclination of the line to the horizontal, and the cohesion is measured as the intercept of the line on the vertical axis.

Shear strength is a time dependent quantity; it may be significantly affected by the rate of loading. When greater loading increments are applied or the rate of strain is increased, greater strengths can be obtained, particularly for cohesive soils. Shear strength may also be influenced by the deformation state. Accordingly, an ultimate or peak value of shear strength can often be distinguished (in analogy to the static/dynamic friction coefficient).

1.2.2.2 Compressive and tensile strength testing of soils

An important and frequently measured physical property of soils is compressive strength. From this quantity, one can determine not only the compressive bearing capacity of deformation characteristics of the soil but its shear strength as well. In the practice of soil mechanics, various kinds of testing methods—unconfined, laterally confined and triaxial compression tests—are accepted. This order corresponds to their historic development. The principle of these tests and interpretation of the models is illustrated in (Fig. 1–13). The unconfined compression test is a simple laboratory procedure (Fig. 1–13*a*). It is performed by using a cylindrical sample usually taken from bore-holes and having a diameter/height ratio of 1/1.5 or 1/2. In test, the sample is loaded between two flat plates and during the test, elapsed time, axial displacement and the sequentially increased forces are measured. After plotting axial strain vs. applied force, the failure stress is determined. The sample deforms not only in the axial direction, but it bulges out laterally when it is in the plastic state. Samples taken from less plastic clays are sheared along an inclined sliding surface at failure. Inclination of the failed sample can be measured and closely approximates the value verifiable by calculation as $\theta = 45° + \varphi/2$. From this value, the internal friction angle φ can be calculated more or less precisely.

Boundary disturbances on the perimeter of the loading plates originating from surface friction cause errors in the value of failure load by constraining deformations on both ends of sample. The central-symmetrical horizontal plane deforms largely without restraint which explains why bulging occurs. The disadvantage of the unconfined compression test is that only cohesive samples can be tested. Granular soils do not have unconfined compression strength.

The result of an unconfined compression test can be represented by a *Mohr* circle whose centre lies on the horizontal axis at a distance of $\sigma_1/2$ from the origin and which is tangential to the vertical axis (τ). The diameter of the *Mohr* circle is the axial principal stress σ_1, which at failure is the failure stress q_u. In soft clays, the internal friction angle is approximately zero, so $c = q_u/2$.

For transitional soils the unconfined compression strength can be estimated by using the empirical relationship

$$\log q_{\mathrm{u}} = as + bv + cl \tag{1-4}$$

where s, v and l values are the relative volumes of solids, water and air, respectively, introduced earlier, and a, b and c values are constants (Fig. 1–16; *Kézdi, 1979*).

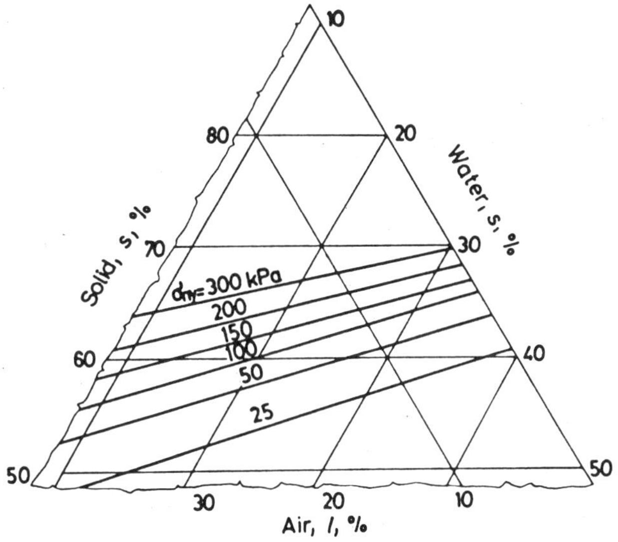

Figure 1–16 Unconfined compression strength for transitional soils as a function of phase composition (*Kézdi, 1979*)

The unconfined compression strength of clays can be correlated directly with the consistency index values shown in Table 1–5. The unconfined compression test is greatly influenced by the size and shape of the sample. The unconfined compressive strength should therefore be considered only as an index property to characterize the consistency of the soil.

In the confined compression test (also called the oedometer test) the axial load is increased gradually while the axial strain, which depends on the elapsed time is registered during the test.

This type of test is mostly used to test soft cohesive soils, which have a small internal friction angle and where the friction of the soil on the internal surface of the confining ring can be neglected.

Table 1–5 Consistency of clays characterized by the value of unconfined compressive strength (*Kézdi, 1977*)

Consistency	I_c	Uniaxial compressive strength kPa
Very soft	≤ 0.25	< 25
Soft	0.25 ... 0.50	25 ... 50
Medium	0.50 ... 0.75	50 ... 100
Hard, rigid	0.75 ... 1.00	100 ... 200
Very rigid	1.00 ... 1.50	200 ... 400
Extremely rigid	> 1.50	> 400

In the confined compression test (Fig. 1–13*b*) the specimens are cylindrical and are loaded by means of two porous disks at both ends (Hungarian Standard MSz. 14045/10–70). This arrangement differs from the unconfined compression test in two respects: the specimen is surrounded by a completely rigid metal ring and therefore lateral strain—bulging—is prevented, and the diameter/thickness ratio of the specimen is about 5 : 1. The soil has no oedometric ultimate bearing capacity (confined compressive strength). As σ_1 approaches infinity, then ε_1 approaches a constant value. In graphical form, a series of *Mohr*-circles — with increasing radii and their centres moving sequentially away from the origin — represent the stress states corresponding to the loading stages in the oedometer test. The point on the right side of the horizontal axis intersected by a circle represents the compressive strength. The point on the left side of the horizontal axis crossed by the circle corresponds to the second principal stress

$$\sigma_2 = (\sigma_3) = \lambda_0 \sigma_1,$$

where λ_0 denotes the earth pressure coefficient at rest. This equation indicates that a soil sample loaded in the oedometer ring never reaches a state of failure and the *Mohr* circles representing the actual loading stages never touch the *Coulomb* envelope. Vertical strains plotted against the pressures give compression curves such as that shown in Fig. 1–17.

From the compression curve the constrained modulus (or modulus of compressibility) can be calculated. This modulus is the ratio of the axial stress to the axial strain, i.e.:

$$E_s = \frac{\Delta\sigma_1}{\Delta\varepsilon_1}.$$

Typical values of E_s for various types of soils in different states are given in Table 1–6. Data in this table illustrate the wide range of soil strength (the ratio of the highest to the lowest values is over 40!).

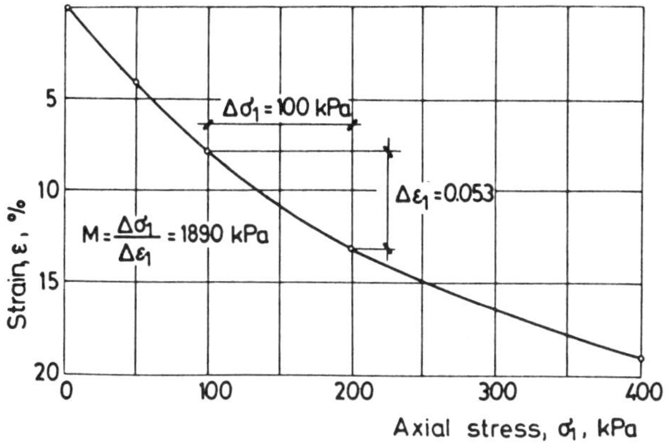

Figure 1–17 Compression curve

Table 1–6 Compression modulus for different types and states of soil (*Rózsa, 1977*)

Soil			Modulus of compression E_s, MPa		
type	name	ρ_d	dry	wet	saturated
Cohesionless	gravel	1.77 1.93	—	3 10	—
	sandy gravel	1.93 2.25	— —	20 70	— —
	sand	1.61 1.93	20 50	15 40	10 40
		1.44 1.92	10 20	7 15	5 10
Cohesive			$I_c = 1.2$	$I_c = 1.0$	$I_c = 0.5$
	silty sand loam	1.57	12...15	8...12	5...7
	clayey sand loam $I_p = 1\ldots7\%$	1.78	22...25	12...15	6...8
	silt, sandy clay $I_p = 7\ldots15\%$	1.35 1.59 1.80	10...15 15...20 18...25	6...9 8...10 9...14	3...6 5...7 6...9
	clay $I_p > 15\%$	1.53 1.72	15...25 25...30	5...10 8...15	2.5...4 5...6

The confined compression test is suitable for measuring not only the structural axial strain caused by rearrangement of solid particles in the soil but also for measuring the strain resulting from the expulsion of water from the voids in the sample. This is the phenomenon that gives the consolidation character to the soil. (This is why this type of test is called the consolidation test.) Knowledge of the consolidation theory of soils is essential in understanding the behaviour of reinforced, soil-made structures. The consolidation behaviour of a cohesive soil is primarily dependent on the saturated or quasi-saturated state of the soil. When saturated, the applied load is initially carried by the water in the pores of the soil. Only after a considerable time will the load be transferred to the soil skeleton. While squeezing the water out of the soil, the neutral stress u decreases which is accompanied by an increase in the effective stress σ. This improves the shearing resistance of the soil. The rate of transfer of stress from the pore water to the solids depends on the type of the soil (mostly on its coefficient of permeability) and on the intensity of the axial load. The rate of loading and boundary drainage conditions also influence the rate of consolidation, e.g. the results are different if drainage can occur both upwards and downwards through the porous stones or drainage is possible in one direction only.

The triaxial compression test is a sophisticated method for testing soils (Fig. 1–13*c*). For this test, a special apparatus has been constructed in which the state of stresses existing in the ground may be reproduced. The triaxial compression test yields useful stress–strain relationships and, also, the shear strength parameters from which the ultimate bearing capacity of soils can be calculated.

In the homogeneous isotropic half-space the soil element is in the 'at rest' state of stress. This means that the vertical geostatic and lateral (horizontal) pressures (σ_1 and $\sigma_2 = \sigma_3$) are principal stresses and the ratio of σ_2 to σ_1 is λ_0 (Fig. 1–18*a*).

As a result of a localised distributed vertical load on the surface, σ_1 increases faster than $\sigma_2 = \sigma_3$ and, for this reason, failure occurs (Fig. 1–18*b*). During excavation, the lateral support of the soil is removed while vertical pressure remains unchanged (Fig. 1–18*c*). By means of a triaxial compression test, any of these stress states can be modelled. The main part of the triaxial apparatus is the pressure cell into which the cylindrical soil samples can be positioned. The sample is protected by a rubber membrane which is surrounded by water under hydrostatic pressure in the first stage of the test. In the next stage, the sample is loaded in the vertical direction similarly to the unconfined compression test. By superimposing the axial pressure and the hydrostatic state of stress any arbitrary stress path can be brought about. The various stages of loading can be represented by *Mohr*-circles as shown in Fig. 1–13*c*. The geostatic pressure condition existing in the soil can be modelled by bringing the specimen into a state of stress where $\sigma_2 = \sigma_3 =$ constant and σ_1 is gradually increased over σ_2. This situation is represented by the left hand side point of intersection on the abscissa remaining at the same

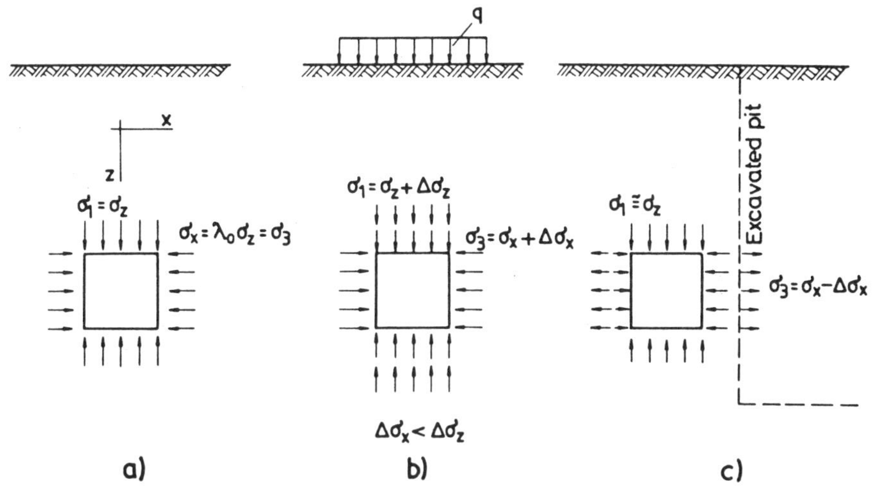

Figure 1–18 Changes in the state of stress of a soil element

a) in the state 'at rest'; *b)* under a surface load q; *c)* close to the wall of an excavation; *d)* continuous load acting on the surface

place while the right one moves to the right during each load increment. The state of stress in the soil adjacent to the side of an excavation is modelled by a test in which σ_1 is maintained constant. This situation can be represented by *Mohr*-circles whose left hand side intersection point with the horizontal axis on the left hand side moves leftwards towards the 0-point. Failure is reached when the *Mohr*-circle becomes tangential to *Coulomb*'s envelope. The specimen deforms in the same way as it does in the unconfined compression test: the rigid specimens exhibit brittle failure and the soft ones exhibit plastic failure, preceded by bulging.

Triaxial compression is a useful test for determining additional soil properties. With advanced equipment the lateral strains (bulging), change in volume and pore pressures can be measured. The initial value of pore pressure as well as the lateral stresses can be set to pre-determined levels. Also, a strain rate can be selected which allows a more realistic approximation of actual stress conditions. In these triaxial compression tests not only the internal friction angle and cohesion can be determined. By introducing some simplifying assumptions regarding the stress–strain conditions within the specimen, other important quantities, e.g. secant or tangent moduli, *Poisson*'s ratio, earth pressure at rest and consolidation parameters, etc., can also be deduced. Concerning the errors caused by the boundary conditions, these problems are similar to those discussed in connection with the unconfined compression test. Friction along the two loading plates increases the strength of the sample and disturbs the strain field. Efforts to eliminate or to decrease the errors are similar, e.g. lubrication of the loading plates, increasing the dimensions of the specimen, etc.

Tensile strength in soils is very small. In stiff cohesive soils the tensile strength can be determined simply by splitting a horizontal cylindrical sample between two loading plates by using the same loading apparatus as in the unconfined compression test. The specimen is then loaded vertically along its two opposite sides parallel to its axis (Fig. 1–13*d*). This laboratory test is mostly used for testing stabilized soil samples. Realistic results in this testing method are obtained only if the soil is at least 'stiff' and failure occurs along a burst rift. During the test and at failure a compound state of stress exists within the specimen and the *Mohr*-circle representing the state of failure is centred around the 0-point of the coordinate system.

1.3 EARTH PRESSURE

Gravity induces stress in earth constructions (self weight stresses). The tensor field is indeterminate, only the vertical stress can be defined in simple cases. For a homogeneous half-space (Fig. 1–19) the vertical stress is:

$$\sigma_z = h\gamma \qquad (1\text{–}5)$$

Horizontal stresses σ_x and σ_y can be determined only if additional suppositions relating to deformation and further material characteristics are introduced in the calculation.

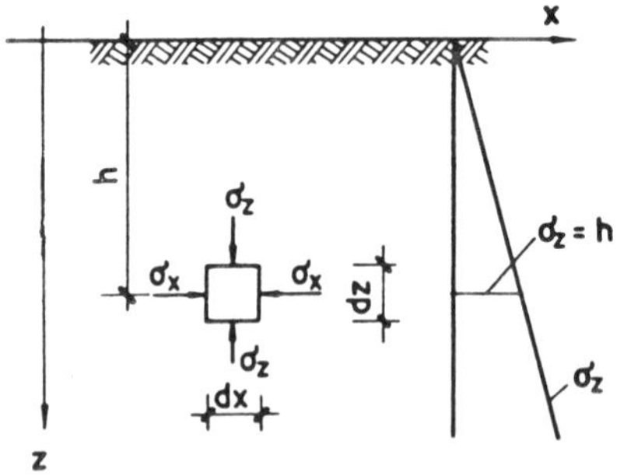

Figure 1–19 Stress conditions in a half-space in rest

γ = unit weight of soil

For example, in the case of a homogeneous, isotropic, linearly elastic half-space:

$$\sigma_x = \sigma_y = \sigma_z \frac{\nu}{1-\nu} = \lambda_0 \sigma_z \qquad (1\text{–}6)$$

where ν is *Poisson*'s ratio, and λ_0 is the coefficient of earth pressure in rest.

Laboratory or field determination of ν for soils is problematic.

In many cases, (e.g. along the interface between a soil mass and a retaining wall) the direction of principal stresses does not coincide with those assumed theoretically. Therefore, the use of relations (1–5) and (1–6) is limited and it is better to replace elasticity theory by the failure stress theory when earth pressure is calculated. For cohesionless soil under half-space (unloaded) conditions, *Rankine*'s theory (1860) is valid if failure occurs according to *Coulomb*'s equation (Sect. 1.2.2.1).

Under these conditions, two limiting equilibrium states (active and passive) can be calculated for the vertical stress σ_z (Fig. 1–20):

$$\sigma'_x = \frac{1-\sin\varphi}{1+\sin\varphi}\sigma_z = \tan^2\left(45^\circ - \frac{\varphi}{2}\right)\sigma_z = \lambda_a\sigma_z; \qquad (1\text{–}7)$$

$$\sigma''_x = \frac{1}{\lambda_a}\sigma_z = \lambda_p\sigma_z \qquad (1\text{–}8)$$

where λ_a and λ_p are the coefficients of active and passive earth pressure, respectively, and φ is the angle of internal friction of the soil.

The lower value σ'_x refers to the condition in which uniform horizontal expansion prevails in the earth mass. The inclination of the potential sliding surface is $45^\circ \pm \varphi/2$.

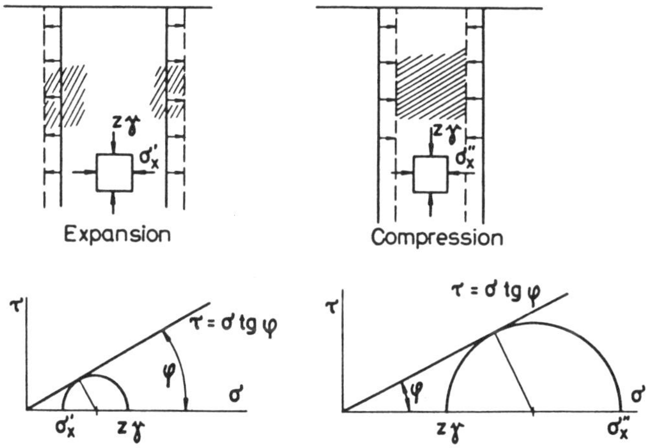

Figure 1–20 Active and passive *Rankine* states and pertinent *Mohr* circles

The magnitudes of the expansive and compressive strains causing failure cannot be calculated theoretically. Figure 1–21 illustrates that the

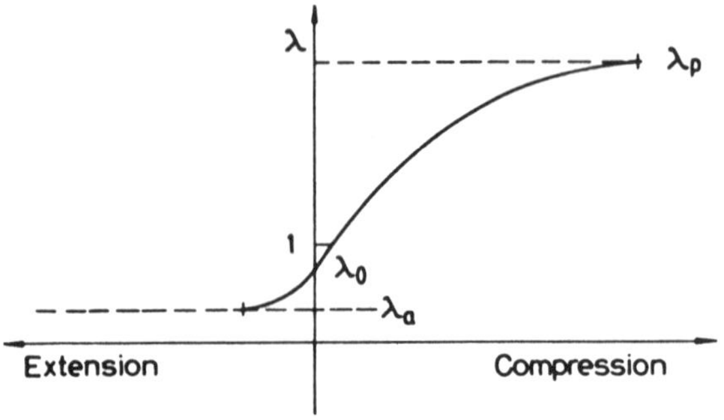

Figure 1–21 Correlation between coefficients of earth pressure and horizontal strain

state of rest (the initial condition) is somewhere between the two limiting states. Triaxial tests indicate that the value of λ_0 depends, among other factors, on density, moisture content and degree of preconsolidation.

Some characteristic values are given in Tables 1–7 and 1–8. An acceptable value can be calculated from *Jáky*'s approximation:

$$\lambda_0 = 1 - \sin\varphi.$$

Table 1–7 Coefficients of earth pressure at rest (laboratory testing data) (*Bishop, 1957, 1958* and *Simons, 1958*)

Type of soil	w_L	w_P	I_p	λ_0
Loose, saturated sand	—	—	—	0.46
Dense, saturated sand	—	—	—	0.36
Compacted residual clay	—	—	9.3	0.42
Compacted residual clay	—	—	31	0.66
Undisturbed, organic silty clay	74.0	28.6	45.4	0.57
Disturbed kaolin	61	38	23	0.66
Undisturbed seabed clay, Oslo	37	21	16	0.48
'Quick clay'	34	24	10	0.52

If the coefficient of earth pressure is calculated for a retaining wall of unlimited length, inclined back wall on which friction resistance is assumed, and inclined backfill (Fig. 1–22), the result becomes rather complicated. *Kézdi (1977)* derived for the active state:

$$\lambda_a = \frac{1}{\sin^2\beta}\left[\frac{\sin(\beta-\varphi)}{\sqrt{\sin(\beta+\delta)} + \sqrt{\frac{\sin(\delta+\varphi)\sin(\varphi-\varepsilon)}{\sin(\beta-\varepsilon)}}}\right] \tag{1-9}$$

where ε is the inclination of the backfill; β is the angle of the back wall to the horizontal; δ is the angle of friction on the back wall.

Table 1–8 Laboratory testing results for coefficients of earth pressure at rest

Soil state	Void ratio e	Coefficient of pressure at rest λ_0
Dense	0.60	0.49
Medium	0.70	0.52
Loose	0.88	0.64

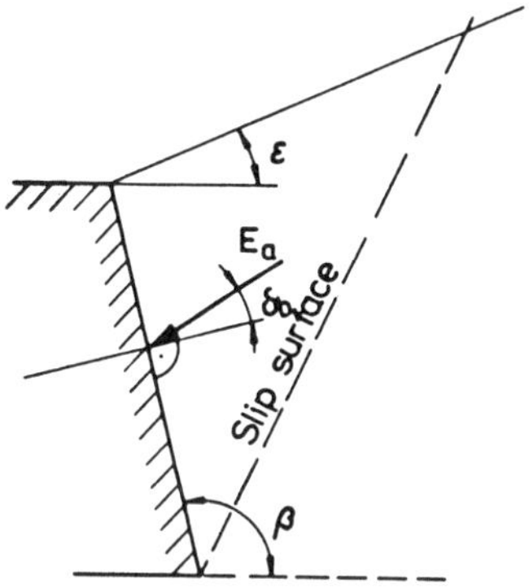

Figure 1–22 The case of inclined back wall and inclined backfill

Similarly complicated expressions characterize the condition where the soil possesses cohesion (c) in the initial state and where the failure condition is described by the expression

$$\tau = \sigma \tan \varphi + c \tag{1–10}$$

The force acting on the vertical back wall (*Kézdi, 1977*), omitting the analysis of stress distribution but assuming linear variation of vertical stress σ_2, can be obtained from the expressions:

$$\begin{aligned} E_a &= \frac{h^2\gamma}{2}\tan^2\left(45° - \frac{\varphi}{2}\right) - 2ch\tan\left(45° - \frac{\varphi}{2}\right), \\ E_p &= \frac{h^2\gamma}{2}\tan^2\left(45° + \frac{\varphi}{2}\right) - 2ch\tan\left(45° + \frac{\varphi}{2}\right). \end{aligned} \tag{1–11}$$

From an analysis of these equations, some practical conclusions can be drawn. Equation (1–9) shows, for example, that the value of λ_a

depends mainly on the value of the internal friction of the soil (φ), whilst the inclination of the backfill has more influence than the inclination of the back wall (*Kovácsházy, 1966*). Notwithstanding the validity of these equations, their accuracy might be questioned because conditions in practice deviate from the theoretical assumptions. So, for example, retaining walls suffer tilting as well as displacement and therefore the expansion (or compression) of the earth mass is far from being uniform as has been assumed.

Sometimes, silo pressure may develop in the earth mass leaning towards the wall. Pertinent results about the spatial force distribution can be derived from Fig. 1–23 as discussed below.

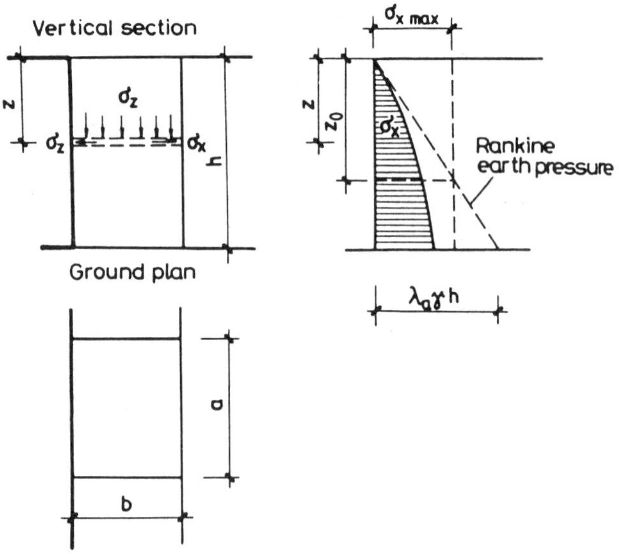

Figure 1–23 Spatial silo pressures (*Kovácsházy, 1966*)

If the side lengths of an elementary cell are a and b, with area A and periphery K, then:

$$A = ab; \qquad K = 2(a + b)$$

and

$$\frac{A}{K} = \frac{ab}{2(a+b)} = \frac{m}{2(m+1)}b, \qquad (1\text{–}12)$$

where $m = a/b$ is the shape factor. Let δ be the friction angle of the cell's surface and $\overline{\varphi}$ be the internal friction of the filling mass, then by using the equation: $\lambda_a = \tan^2\left(45° - \frac{\overline{\varphi}}{2}\right)$, we obtain the vertical stress

(*Kovácsházy, 1966*):

$$\sigma_z = \frac{A\gamma}{K\lambda_a \tan\delta}\left(1 - e^{-\frac{K}{A}\lambda_a z \tan\delta}\right) \tag{1-13}$$

and for the side pressure:

$$\sigma_x = \lambda_a \sigma_z .$$

The influence of silo pressure is substantial wherever $m \leq 1$. The variation in the stresses in direction z is determined by the exponential term. If the critical depth is

$$z_0 = \frac{A}{K\lambda_a \tan\delta} \tag{1-14}$$

then the horizontal stress is:

$$\sigma_x = \gamma z_0 \lambda_a \left(1 - e^{-\frac{z}{z_0}}\right) \tag{1-15}$$

From this relation, at depth $z = z_0$, we find that the silo pressure is one-third of the active pressure. It is worthwhile to recall this fact when earth pressures are calculated during retaining wall design.

Designing an earth-made structure requires considerable care and one of the most important steps in the design process is the determination of the earth pressure. Most countries have elaborate Standards and Codes of Practices for this purpose. Emphasis should be laid on the displacement and deformation of the retaining structure when earth pressure is calculated. Also, the interaction between earth mass and wall should be investigated. This might be more pronounced in the case of earth-made structures. Therefore, emphasis is given in this book on methods of calculating earth pressures.

2

METHODS FOR IMPROVING SOIL PROPERTIES

2.1 THEORETICAL CONSIDERATIONS

In section 1.2, the mechanical properties that dominate soil strength were discussed. By referring to the relevant information, we intend to survey how these properties may be improved in what follows. The strength of soil is largely influenced by the character of the bonds existing between the soil particles. For simplicity, it is often implicitly assumed in calculation that soils are weightless and stressfree and the quality of interparticle bonds is expressed in terms of material parameters. This is, however, not true in the physical sense because the state stresses cannot be disregarded even if the mathematical model gives a good approximation to reality.

Among the variety of bonding mechanisms, the most simple is that caused by gravitational forces. Elementary reactive forces act at contact points between the grains even where other physical or chemical bonds are negligible (i. e. among pebbles). The sum of the contact forces on the surface of an individual grain balances the dead weight; but the former might produce stresses which are very high in comparison to the latter, due to the Hertzian nature of the contacts.

In all other soils except gravel (and perhaps a few types of sand) other bonds exist between the grains forming a stable skeleton. Moisture content and mineral composition play the predominant role here. *Kézdi* (*1977*) and *Mitchell* (*1976*) emphasize in their monographs the substantial effect of the liquid component on strength in describing small changes in moisture content which result in alteration in the mechanical properties of soils. The sensitivity of plastic soils is well know in practice but fine graded soils may also exhibit similar properties.

As a result of various mechanical, physical and chemical bonds, pressure forces initiate shear force components in the tangential plane at the contact points between two particles. Therefore, the direction of the resultant force deviates from the normal to that plane. All of these individual shear components lend shear strength to the soil mass. So it can be seen that the shear strength, which characterizes the mechanical behaviour of a soil mass, originates as a result of bonding forces and is not due to stresses between the particles.

Additional external loads and effects may greatly modify the initial pressure and shear stress in a soil. This causes deformation of the original skeleton. Changes in the magnitude of bonds (activation of new bonding forces, dilution of bonds by increased moisture content, etc.), due to several miscellaneous sources, contribute to changes to the pattern of deformation. With the exception of some particular clay minerals, deformation of the shape of a particle is far less than the displacement of particles in relation to each other. Modulus of elasticity tests in the laboratory or in-situ reveal slip and angular displacement of particles.

Considerable compressive and shear strength may develop in a soil but the solid skeleton is hardly able to bear any tensile stresses. This deficiency is the principal justification for improving soil characteristics. Two ways are open to improve tensile strength in soils:

— first, to enhance bonding energy between particles,
— second, to incorporate into the soil mass foreign materials of high tensile strength.

Mechanical, physical or chemical interactive processes may help to increase the level of bonding energy between particles.

Success in improving soil strength by incorporating foreign materials of high tensile resistance into the soil depends mainly on the interaction between inclusion and soil. Friction and adhesion are the most important issues in this respect but other factors influencing the deformation of the inclusion and the surrounding soil should also be considered. Numerous materials are fit for application. Our review classifies applicable processes according to the type of inclusions, their geometric shape and their distribution in the mass.

Where earth-made retaining walls are considered, the final choice will depend on the effective use of reinforcement and efficient use of resources. Detailed information will be given in what follows about favourable applications. A few cases of limited applicability will also be discussed which should also prove useful to the reader.

2.2 PRACTICAL METHODS

2.2.1 Compaction

A well-known mechanical method for altering soil condition is to modify the phase relations (ratio of solids to liquids and to air). Compaction increases the mass of solids per unit volume thereby increasing the number of contact points. Force transfer is intensified and greater stability is created in the soil structure. With regard to the variety of soil types, the methods of compaction are diversified. Besides rolling, tamping and vibrating, other methods like blasting or drop hammers are often encountered in practice.

Correlations between density (characterized by parameters e, ρ_d, and T_{re}) and shear resistance τ are shown in Fig. 2–1 (*Ryzhov, 1975*). Bearing capacity and stability of an earthwork is largely influenced by mass density. Typical values for the modulus of elasticity of various soils can be seen in Table 1–6 where the importance of different soil conditions is also indicated.

Compaction intensifies force transfer between particles and induces a new stress distribution pattern and increased shear strength. This condition differs widely from the initial state and is not always favourable because, under special circumstances, an extremely dense earth mass may become rigid to an unwanted degree. Transferred energies will not be dissipated by internal deformation, whilst displacement of the rigid mass will take place as one unit. In the case of earthworks, this process could entail a loss in stability.

Phase relations are also altered by changing moisture content in a way parallel to compaction, and involves physical and chemical bonding effects on the interface of particles. Shear strength varies nonlinearly with change in water content because bonding energy levels are primarily influenced by physical effects on the surface of particles. For greatest efficiency, clays usually need drying but dry silts need wetting. A well-known phenomenon is the so-called 'apparent cohesion' of fine soils. Specifications in some countries allow for this type of cohesion where permanent changes in moisture contents can be maintained. This is, however, dangerous as can be seen from the collapse of sand-pit walls.

Compaction and compaction control is a well-documented field in the literature (*Kézdi, 1977*; *Rózsa, 1977*; ME 19–23). Here, only a comprehensive summary of the conclusions influencing the construction of earth-made retaining walls is given:

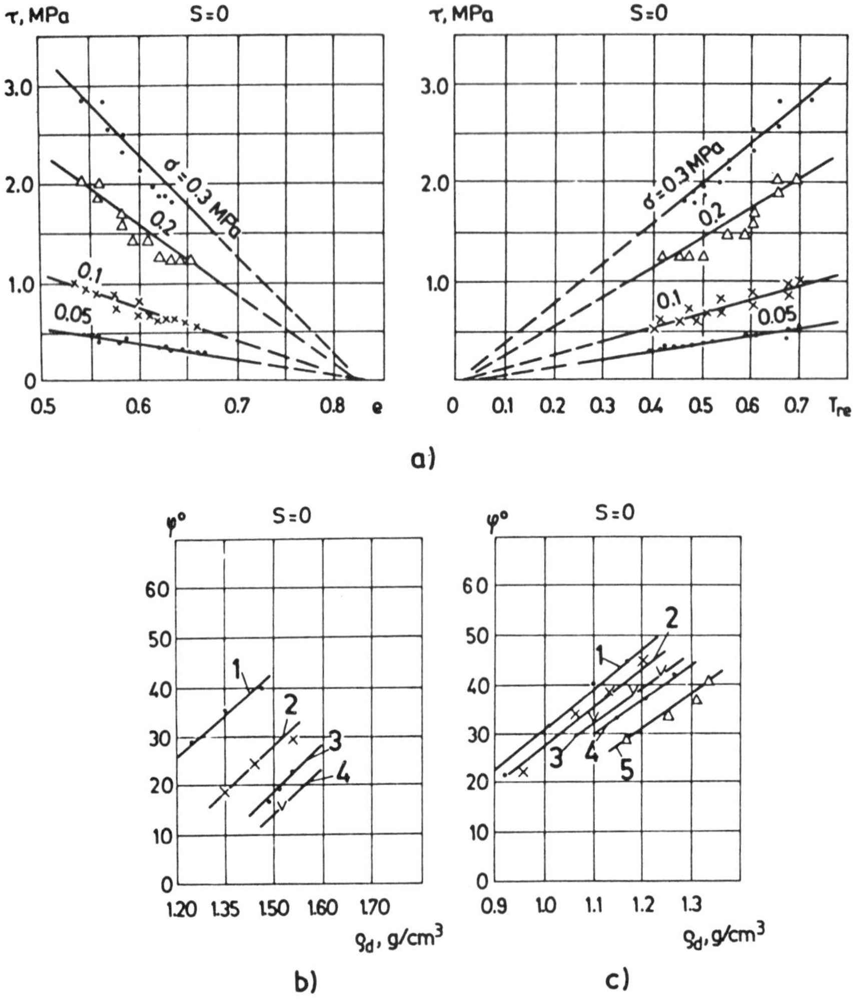

Figure 2–1 Relationships between strength and state parameters (*Ryzhov, 1976*)
a) medium sand; *b)* loess;
1 5 kPa; *2* 50 kPa; *3* 100 kPa; *4* 500 kPa vertical stress;
c) clay; *1* 25 kPa; *2* 50 kPa; *3* 100 kPa; *4* 200 kPa; *5* 500 kPa vertical stress;
φ friction angle; e void ratio; τ shear stress; T_{re} relative density

— increased density results in improved soil properties; therefore it is essential in the design of earthworks to specify the desired density and to control it during construction;
— in stabilizing earthworks by using inclusions (Section 2.2.5), an essential requirement is to ensure bonding between the inclusions and

the soil mass; therefore careful compaction is required in the zone of inclusions;

— depending on the compaction method, inhomogeneous and anisotropic stress conditions in the earth mass may be induced; therefore, only compaction methods which do not disturb the desired (designed) stress distribution should be employed.
— density may change later due to applied surface loading or dynamic effects; this should be kept in mind in the design of earth-made structures.

2.2.2 Chemical soil stabilization

Bearing capacity of soils can be enhanced and permanently maintained in most cases by injecting substances (mostly in the form of solutions) into the pores of the soil skeleton. Partially or fully filled pores form an obstruction against seepage, infiltration and groundwater movement and lend stability to the treated mass. Some chemicals are formulated to extend and stabilize interactive bonds on the surface of particles. This process is most active in clays. Chemicals that are able to expel water from pores and surfaces of particles may achieve permanent bonding effects by filling the pores completely. Two terms are used in literature (*Kézdi, 1976*):

— to strengthen means to increase soil strength (by using whatever process) without disturbing the soil structure (skeleton) and,
— to stabilize means a process in which soils are disturbed, kneaded or recompacted.

The first procedure is mainly applied in underground construction and near foundations where the surrounding soil needs strengthening. Stabilization has been used in road, railway, runway and waterway construction for decades.

With respect to earth-made retaining structures, both strengthening and stabilization may find applications using local soil, cheap bonding agents and simple technologies to produce gravity walls or to manufacture prefabricated elements. Also, surface drainage, prevention of seepage and peeling can be accomplished by stabilizing a thin layer on the ground surface. The mechanisms of action of commonly used strengthening and stabilizing agents (cement, lime, bitumen, waterglass, acryl-copolymers, etc.) has been subject of extended research since the 1950's. Formulae and techniques have been developed which assure proper quality and durability. Equal reliability can be expected from marketed chemical products. These are patented and therefore only restricted information is given by the manufacturers about the composition of these materials except for instructions about their usage. Some of these chemical agents consist of surprisingly simple substances, being the result of natural observation; others were transplanted from various technologies (concrete, textiles) while others came into existence after intensive research into bonding. Among recently developed products, we can find some which are effective when used as additives at concentrations as low as one or two percent by volume.

Chemical soil stabilizing methods are still rarely applied in earth wall construction being hardly competitive with other solutions. The abundance of marketed products, simple technologies and control methods make it worthwhile, however, to consider their use in earth-made structures. If the idea is accepted, the following aspects should be considered:

— chemical soil stabilization techniques (mixing and compaction) are not competitive economically when small volumes are involved;
— grouting may leave untreated zones which make consumption of filling and reagent materials uneconomic;
— chemical treatment requires environmental and labour protection;
— chemically treated soil will be deprived from fertility, therefore separate overburden and drainage facilities may have to be constructed.

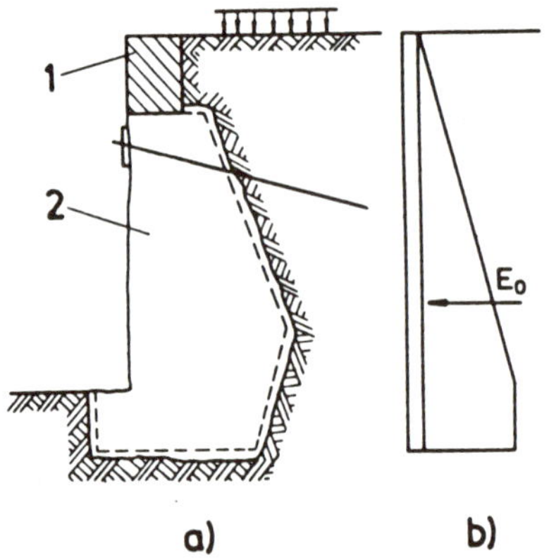

Figure 2–2 Strengthened retaining wall with anchor (sketch)

a) construction; *b)* earth pressure diagram;

1 edge beam; *2* strengthened soil

Despite these discouraging aspects, a trial has been carried out by *Borchert, Müller-Kirchenbauer* (*1981*) as illustrated in Fig. 2–2. Besides conventional stability control, the authors carried out an investigation on the creep of the structure and described a simple manual method for calculation.

2.2.3 Thermal strengthening of soils

Baking of bricks was probably the first thermal treatment of soils, going back to ancient times. Clay and silt (loess) are primarily suitable, but the process is universal. Only one aspect needs attention, namely, the high temperature should not provoke unwanted chemical changes in the mineral content of the soil.

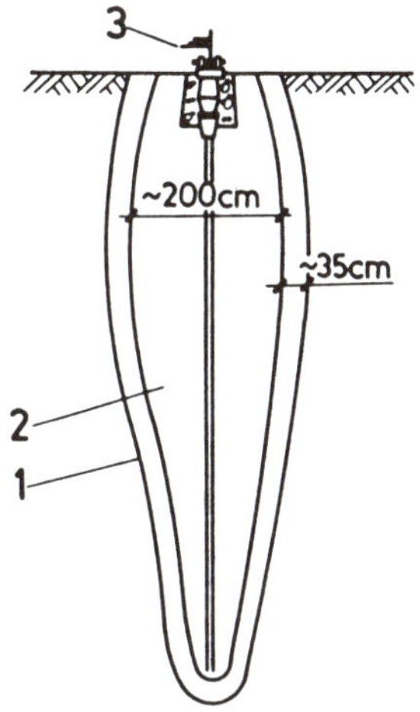

Figure 2–3 Principle of thermal strengthening

1 melted core; *2* strengthened zone; *3* heating head

The principle of the thermal process is shown in Fig. 2–3 (*Rózsa, Kelemen, Regele, 1977*). High temperature should be maintained in a bore-hole driven through the zone to be treated. The surrounding soil becomes heated and partly melts. After cooling, permanent bonds remain between the particles that were lying loose previously. The volume of the strengthened zone depends on the mineral content of soil particles, the thermal conductivity of the soil and the quantity of heat supplied. 1200 to 1600°C has to be reached in the bore-hole to bring the surrounding soil to 300 to 800°C temperature. The lower limit would be usually enough to decrease side pressures substantially, whilst at the upper limit a dense, almost stonelike soil structure is produced that remains resistant even against freezing. (It is worth noting that the temperatures used in the building and ceramic industries are about 900 to 1000°C).

Techniques for thermal treatment of soils require the use of well-established industrial techniques (blowing of furnace-heated air, burning fuel in the bore-hole, soil melting using thermographite heaters, etc.; *Jurdanov, 1978*). These processes require only few workers and their main advantage is the well-mechanized working site. This condition is most valid for recent developments where electric heating is applied. Mechanization makes these processes quick and independent of climatic conditions.

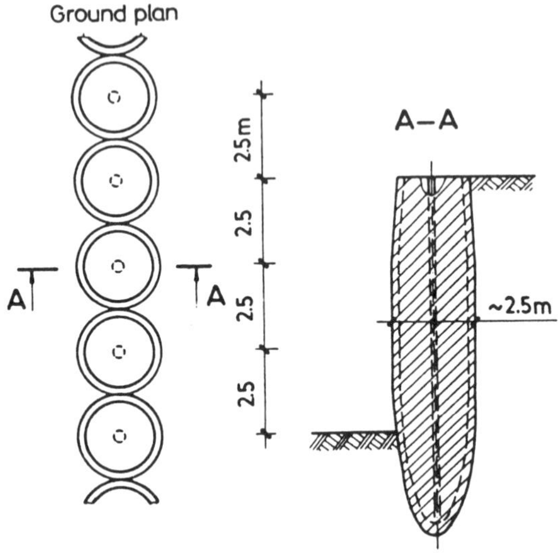

Figure 2–4 Retaining wall of burned soil

Burning produces more or less round columns in lines which may be contiguous (Fig. 2–4), and form a sufficiently resistant retaining structure. Depending on the techniques used, the distance between bore-holes is usually 2 to 4 m and their lengths are 3 to 5 m. The resulting structures are similar to gravity walls. Excavation of superfluous material can be mechanized and face finishing adjusted to demand.

A considerable disadvantage of these methods lies in energy consumption. The aim of the process is to modify soil structure, i.e. to strengthen it, but in a definite zone in an extended medium. Though soil has poor heat conductivity and burning might be short lasting, loss of heat energy is unavoidable. Heat dissipates unnecessarily into the neighbourhood and partly into the heat conducting equipment. Also, losses occur in warming up the treated zone to the desired temperature. Groundwater might also be an obstacle. According to data in the literature (*Jurdanov, 1980*), fuel consumption for burning one m^3 of soil needs about 60 to 80 kg of oil, that is, two-thirds of the quantity used in factories for baking the same volume of clay products. This information gives quantative data for deciding on the applicability of the process. Even if the boring and burning equipment is lightweight, mobile and mechanized, it is hardly probable that this approach would be competitive in comparison with less energy consuming alternatives.

Another process where thermal methods are used for improving soil characteristics is freezing. Detailed discussion of this topic is, however, omitted from this review because this process can be used to produce temporary structures only under normal terrestial climatic conditions. Comprehensive information can be found in the literature (*Rózsa, Kelemen, Regele, 1977*).

For completeness, we mention in passing that in permafrost regions where extended building activities have followed population growth—due to mining and industrialization—research into the mechanical behaviour of frozen soils has been initiated. The literature deals with soil characteristics and offers a variety of structural solutions. Mostly English and Russian books and articles are of interest.

2.2.4 Soil strengthening by using plants

Natural steep hill sides are good examples of how vegetation contributes to stability. This section deals mainly with the role of vegetation in lending strength to the ground. Illustrative examples are described in Section 7. The roots of grass, shrubs and trees form a fabric in the 0.1 to 1 m thick upper layer of ground. In most cases, the top layer in these areas is the result of weathering or sedimentation. Roots are strong fibres or threads, growing in a random pattern and can sustain considerable tension, making the soil woven through by them a composite strong mass. The tearing force of roots can be illustrated by data given by *Bengemann (1978)*:

bearded wheat (*Agropyron repens*)	16 MPa
brittle willow (*Salix fragilis*)	18 MPa
white willow (*Salix elaeagnos*)	16 MPa
lucerne, alfalfa (*Medicago sativa*)	46 MPa.

No data are available about the shear strength of roots but observations suggest that roots may raise the shear resistance of soils by orders of magnitude. Soil strata woven through by roots act like gravity walls and can sustain steep slopes (*Siedek, 1965*). Early environmental investigations showed that some plant roots grow in a peculiar manner in sloping ground (Fig. 2–5). For example, willow-herb (*Epilobium angustifolium*) extends long roots in the opposite direction to the slope and is so able to suspend its stem (*Schiechtel, 1975*).

Figure 2–5 Retaining roots in a sloping soil zone

Another example in Fig. 2–6 shows the steep slope of some earthworks fastened with plant roots. Two metre long, 15 to 20 mm thick willow strips (Salix) were laid above the lifts during construction. In a few months time, 0.1 to 0.2 m long bushy roots grow to the benefit of the earthworks. Grassing and seeding of the intermediate strips on the slope is also most desirable.

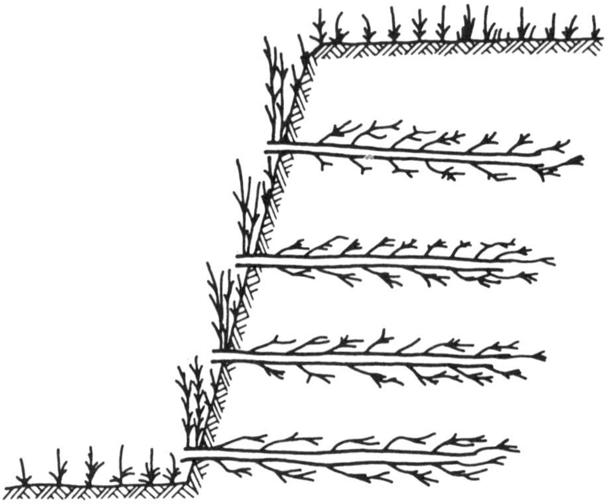

Figure 2–6 Steep hillside strengthened with plant layers

Environmentally protective earthworks strengthened with plants have appeared recently as noise attenuating and sight restricting constructions along motorways. Well established shrubs and trees with dense roots are planted on them, according to the climatic conditions, selected from the following species:

country maple	(*Acer campestre*)
speckled alder	(*Alnus incana*)
silver poplar	(*Populus alba*)

trembling ash (*Populus tremula*)
elder (*Sambucus racemosa*)
privet (*Lingustrum vulgare*)
cornel (*Cornus mas*)
thorn bush (*Hippophae rhamnoides*)

and others: (*Prunus padus, Sorbus aria, Lonicera xylosteum, Vibruum opolus,* etc.).

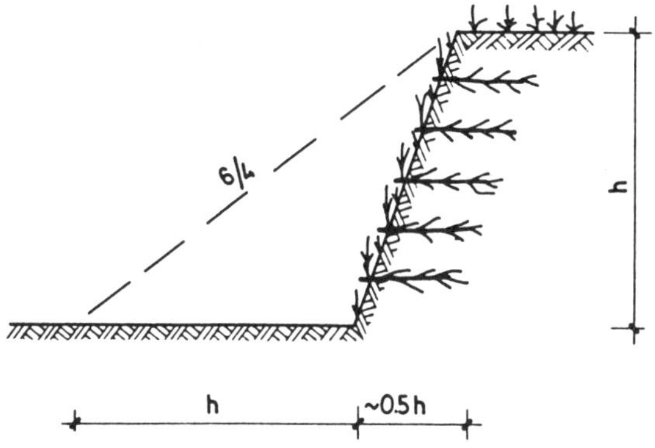

Figure 2–7 Ground surface gained by use of soil wall

Engineering considerations, given the soil conditions and botanical factors dictate the most suitable arrangement for plantation and selection among species. Mineral composition, soil structure and pH-value of the soil should be taken into account (*Schlüter, 1971*). The slope angle of a silty clay embankment, an unfavourable soil in this respect, can be designed steeper where plants are grown. A strip on the ground surface as wide as the wall's height can be saved if the earthwork is constructed in this manner (Fig. 2–7). Acclimatization of properly selected plants might be required in some regions. Swiss experience indicates, e.g., that six weeks aridity in July did not harm *Lonicera xylosium* (*Woodtli, 1978*). Because of the importance of vegetation in earthwork construction, this topic will be returned to again in the following sections.

2.2.5 Soil stabilization by using inclusions

It has already been mentioned in the 'historical background' section that the stress distribution in earth masses can be improved by incorporating foreign materials. Since the aim of this book is to specify appropriate solutions, a separate section will deal with this question. Here, only some general aspects are discussed in order to set the subject into context.

The best known simple case in this respect is adobe. Adobe consists of a silty lean clay to which shortplant fibres are randomly mixed, producing an isotropic mass of improved characteristics after drying. It is therefore evident, in seeking higher strengths, to make use of this popular idea (analogous to steel scrap concrete) by admixing artificially produced fibres or threads to soil (*Szabó, 1976*).

Strengthening by using inclusion is mostly done by laying mats, nets, grids, sheets, etc. horizontally on top of compacted lifts. This simple technique reflects the structure's mechanical state and the subsequent deformation pattern (whereby flat cylindrical surfaces develop) which can be easily followed by the planar inclusions.

A wall system composed of vertically inserted wall elements perpendicular to the plane of the slope (see Fig. 6–22) is also a reinforcing method of this type. In this case, the retained mass is reduced and the side pressures acting on the slabs become more prominent. Its application is advantageous for reinforcing cut slopes; more details about this arrangement are given in Section 6.

A planar and linear reinforcing system of bands and strips (instead of mats) in a horizontal arrangement above the lifts in the earthwork is also used. This was the method that triggered interest in earth reinforcement a few decades ago. Detailed examination of pertinent solutions and construction technology is presented in Section 5.

Rods, wires and flat steel can also be used as inclusions in earth-wall reinforcement. Soil nailing or the use of root-like piles also belong to this group. These solutions are mostly applied to cut slopes where the retained earth mass needs reinforcement but excavation and wall building is not desired. Section 6 sheds more light on relevant solutions to these problems.

2.2.6 Composite strengthening methods

So far an outline review has been given of earth-made structures and methods for reinforcing earth masses. Engineering invention has produced various solutions in practice which are not pure adaptations of one or the other method but combine the features of phase relation modifications and insertion of foreign materials in the soil.

One such solution is, for example, suitable for draining wet soft clays by inserting vertical ribbed plastic laths in the soil to the desired depth. Capillarity changes the phase relations (by collecting, lifting and evaporating moisture) and the laths work as stiffening inclusions. Sand and gravel piles also belong to this group of soil strengthening methods if their arrangement complies with definitions given in Section 1.1. These methods are mainly used where stronger foundations are needed below earthworks to avoid unwanted settlements.

Detailed descriptions of these composite methods are omitted from the following text because their application is rather incidental and their design and performance can be inferred from the descriptions of the parent methods given in the following sections.

2.3 SELECTING THE APPROPRIATE STRENGTHENING METHOD

Alternatives are always considered and evaluated during the design process. In most cases, the question arises whether a conventional gravity retaining wall or an earth-made structure should be constructed. Here, the main factors are compiled to facilitate arrival at a fair decision.

In respect of building materials and energy consumption, earth-made constructions are more economic in comparison with conventional gravity walls because the former makes use of and benefits from mechanical properties of locally available resources. A particular price system may distort an estimate (in some countries), or personal interest may influence decisions, but measured in 'natural resources', i.e. in the real amount of invested goods, the costs are usually much less for earth-made retaining structures than for conventional retaining walls. Obviously, both techniques are costly. Machinery and haulage are major items which counterbalance savings in building materials. Careful cost analyses have to be conducted to reveal the balance of costs, but often it will tilt in favour of earth-made constructions.

In all cases it is also necessary to compose a schedule of works (from site preparation to handing over). Less time is required to construct earth-made structures due to reduced haulage demand and setting time delays in the case of concrete. Maintenance cost and expenditure for corrosion protection are rarely included items in an estimate for conventional gravity retaining walls. These would, however, hardly make less than a similar burden for an earth-made wall. Environmental and aesthetic considerations (to which a price can be allocated) are no drawbacks in the case of earth-made structures where planted surfaces rival any conventional wall face finishing.

Finally, we have to emphasize that one argument, levelled against earth-made structures, needs fair consideration. The construction of earth-made structures should be excluded, for example, from consideration where later displacements or deformations are not permitted but may be expected, but sufficient compaction or time for consolidation cannot be ensured during construction for any reason. There is no need to promote reinforced earth construction techniques by force when, in most cases, their obvious attributes make their own case.

3

REINFORCING SOIL BY USING INCLUSIONS

3.1 PRINCIPLES OF REINFORCING

Effective soil reinforcement depends on efficient matching of the structural arrangement, the loading conditions and the strength of the reinforcing elements. In general, the tensile and shear strength of the reinforcing materials increase the bearing capacity of the soil mass, and decrease the strains in the earth construction, but there are important additional conditions to be satisfied.

There is an abundant choice of soil reinforcing structural arrangements and minor modifications made to satisfy various local requirements further increase the number of possibilities. There are more than a dozen common types and the number of reported innovations grows year by year (*Bartos, 1979 John, 1979*). Classification is therefore necessary to survey the field. The first plausible category (based on the geometry of the reinforcement) was described in Section 2.2. Another category can be defined by the construction technology:

— backfill techniques do not require special machines, the shape of the earthwork determines the technology and the inclusions can be placed during any appropriate phase of construction.

— in the insertion process the initial state of the wall has to be kept undisturbed as far as possible, and special machinery must be used to place the reinforcing elements.

Among the construction methods using backfill techniques, reinforced earth seems to be the most important one. This type of construction has been well developed; design methods and dimensioning details are well-known and national Codes of Practice and Design have been released in many countries over the world. Steel plates or strips (of stainless steel or galvanised steel) are used most frequently as reinforcement where friction forces are exploited at the soil–metal interface. An earthwork interwoven with reinforcing steel strips behaves as a gravity wall.

Materials other than steel strips can be applied, as well. First of all, woven or sewn plastic sheets, extruded grids or fabrics are very popular. Another sort of reinforcement in the backfilling category is shown in Fig. 3–1. Rigid inclusions placed along the sewer pipe restrain soil deformation; this kinematic constraint improves bedding conditions around the pipe (*Magyarné, Scharle, Szalatkay, 1979*).

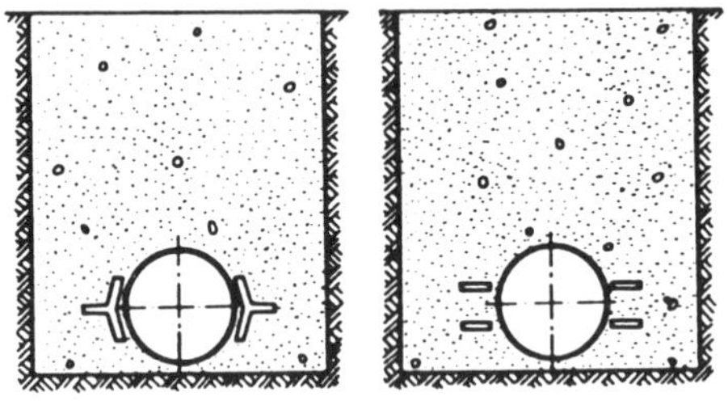

Figure 3–1 Rigid reinforcing elements placed around sewer pipes

The most common ways of soil reinforcement in the insertion category are root piling and soil nailing. Early descriptions of these soil reinforcing techniques can be found first of all in the Italian and German literature. The inclusions (concrete or metal rods of small diameter) are introduced into the undisturbed soil by injecting, trenching, boring, etc. and increase the tensile and shear strength of the soil mass. Accidentally, they may have some direct role in distributing the internal forces which develop under external loads, but in these cases they should be considered foundations rather than reinforcements. Further examples will be presented in Section 6.

For dimensioning reinforced soil structures, with respect to their mechanical behaviour, two questions of great importance must be considered:

- interaction between a reinforcing element and the neighbouring soil, i.e., the force transfer conditions,
- relationship between the strain characteristics of the reinforcing material and the soil.

Theoretical considerations and experimental results show that both questions are rather difficult to answer. Therefore, one must not neglect the rules of thumb inherited from the early period of earth construction and still often used by designers. Many of them are accepted as valid today.

Though earth walls designed in accordance with these crude approximations were cost-effective, the shortcomings of these purely practical, trial-and-error approaches are also evident. These crude approximations

- — do not reflect the genuine mechanical behaviour of a structure, therefore the efforts to improve the computational methods have been fruitless,
- — do not describe the proper kinematics of the construction under unusual loading circumstances,
- — do not yield the actual safety of the construction, so they may be highly uneconomic due to superfluous and hidden reserves in safety.

To illustrate the first point, we refer to the anchoring approach, used very often for dimensioning a reinforced earth wall. Certainly, this method is generally accepted (we describe the approach in detail in Section 5), but the earth mass, densely reinforced with strips of high tensile strength, should be considered as an anisotropic earth block (gravity wall). That is why we do not think that sophisticated numerical techniques can help in obtaining better results with these models.

To illustrate the third point, as a historical example, consider the case of the first reinforced soil wall built in Scotland. The coefficient of safety against surface friction failure along the interface of the reinforcing strip and the soil was determined in two ways. The French and the British methods gave values of 2.4 and 0.7, respectively. The wall was stable, proving that the latter method was 'rather conservative' (*John, 1979*).

The mechanical behaviour of an earth construction strongly depends on the relative straining characteristics of the interacting soil region and the reinforcing elements. Taking into account both theoretical considerations and experimental results, two limits of kinematic response can be distinguished :

- — in the inclusions, which are relatively rigid in comparison with the soil in their neighbourhood, high tensile stresses may develop despite small soil strains. The frictional resistance of the soil–reinforcement interface results in secondary soil pressure. But the bearing capacity is reached at a definite strain level, even it is related to rupture of reinforcement. Earth walls reinforced with metal strips often behave in this way.
- — the reinforcing effect evolves with time if the reinforcement is relatively flexible and deforms with the soil. Large displacements may occur in these cases and it is impossible to define bearing capacity; the limit state can be stipulated by some bound to strain or displacement. Earth constructions reinforced with plastic grids or fabrics behave in this way very often.

It is clear, therefore, that the layout of the reinforcement is of great importance. The most effective arrangement can be reached by placing the reinforcement in line with principal tensile trajectories. Of course, this theoretical optimum cannot be realized because we do not know, as a rule, the exact pattern of the trajectories. Moreover, practical considerations often dictate (or at least, suggest) some regularity in placing. Nevertheless, it is very important to meet this principle since its neglect may result in poor efficiency. Reinforcing inclusions placed along the pressure trajectories or sheared surfaces may decrease the shear resistance of the whole soil mass, if the angle of internal friction of the soil is higher than the friction angle of the soil–reinforcement interface.

Most of the principles mentioned above are plausible. Nevertheless, there are examples where it becomes clear that more than one theoretical principle might be applied to a situation in practice, and only a careful analysis can decide between the efficiencies of alternative arrangements.

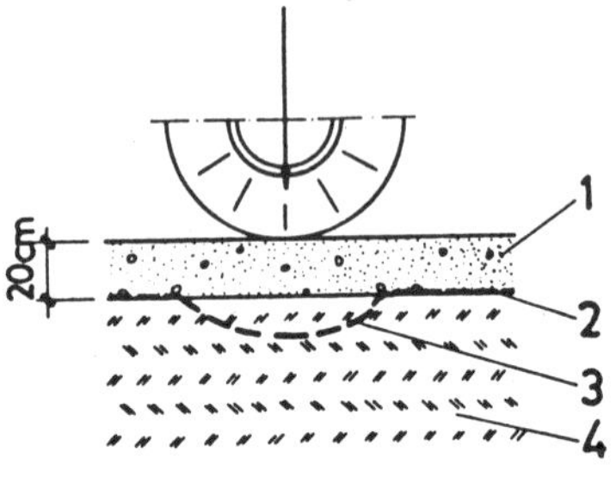

Figure 3–2 Single reinforcing sheet in a road base

1 sandy gravel; *2* reinforcing plastic grid; *3* deformed grid; *4* soft subgrade

Consider for example, the case of a road base reinforced with a plastic grid or textile sheet (Fig. 3–2). This arrangement had been investigated by several authors in the mid-1970's. There were those who said that the results were positive (*Jessberger, 1977*). Other authors, however, reported controversial results, some argued over the magnitude of the reinforcing effect; others questioned even its existence (*Barvasov, 1977, Sorlie, 1977*). In another trial, the beneficial effect of the plastic reinforcing mat placed above the sewer pipes was proved experimentally (*Pálossy, 1979*).

To understand and explain the difference in the results of various authors one must reconsider the fundamental question: whether the strength of the reinforcing material has been mobilized under the working loads or not. In the expansible plastic sheet, considerable strains have to develop as a result of large shear stresses transmitted by the soil, before their reinforcing effect is manifested. Therefore, expansible plastic sheets placed near to the surface may prove to be less effective than those lying at some depth. On the other hand, stiff or rigid inclusions restrain the kinematic degree of freedom between granules (where interlocking and frictional resistance are interrelated) at a low stress level.

Observation of other cases shows that expectations connected with the static and kinematic behaviour of earth-made reinforced structures may turn out to be faulty if the analysis was based (even tacitly) on the axioms of linear elasticity and the assumptions of homogeneity and isotropy. In the following part of this section, the most important results and data pertaining to earth walls are described in some detail.

3.2 SOIL REINFORCEMENT WITH RANDOMLY DISTRIBUTED ELEMENTS

In several examples mentioned in the 'historical background' section, randomly distributed fibres or threads served as reinforcements. The practical application of these variants is strongly bound to construction technology. Since the late 1960's efforts have been made from time to time to find an appropriate field for their use but the results remain unconvincing.

Theoretically, and under laboratory conditions, the reinforcing effect of randomly distributed inclusions is clear. The strength of soils interspersed with fibres increases and their stiffnesses improve. However, to implement these techniques in masses at the construction site is by no means a simple and inexpensive task (*Hoare, 1979*). Furthermore, the soil–fibre mixture must be compacted, as a rule, and this is laborious. The required compactive effort may be much more than that for soil alone; in some experiments the fibre reinforced earth behaved like clay. The economics of reinforced earth structures may be unattractive if the construction process uses up the savings gained in the structure.

For these reasons random reinforcement seems to be an academic idea of little practical importance from the point of view of earth walls. Nevertheless, the idea itself must not be written off, and some recent developments may be worthy of consideration (*Khay, Morel, Perrier, 1986*).

3.3 LAMINATED SOIL REINFORCEMENT

We have mentioned before that the mechanical behaviour of reinforcement placed in layers is well suited to the geometry and construction practice employed in most common earth-made retaining structures. In design and calculation we very often assume two-dimensional sections. In these cases, the slightly curved or plane reinforcing layers appear as lines in the cross section. Their position can be easily compared with the pattern of trajectories determined from continuum theories.

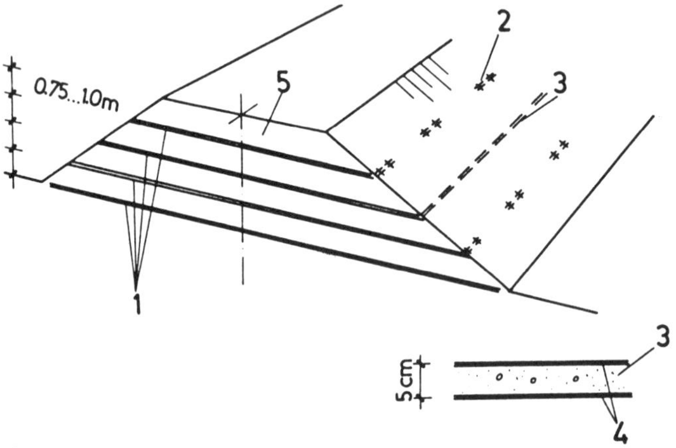

Figure 3–3 Soil embankment reinforced with lime layers (*Yamanouchi, 1982*)

1 lime layers of strips covered with geotextiles; *2* lime strip; *3* lime layer; *4* geotextile; *5* clay

The concept of a reinforcing layer is not restricted to woven, sewn or rolled textiles or grids. We refer here to an interesting example where the reinforcement consisted of dry lime powder (Fig. 3–3). Five cm thick lime layers, packed in geotextiles, were placed in an embankment at distances of 0.75...1.00 m above each other. They improved the stability of the earthwork which was made from a wet local soil of poor characteristics (*Yamanouchi, 1982*). This is a good example to show how useful it can be to apply imagination to general principles. In analysing the reinforcing effect, one must not restrict the investigation to simple mechanical interaction (although in the introductory Chapter we have done so for the sake of simplicity). Due to the dry lime layers, for example,

- — swelling pressures arise which accelerate the consolidation process,
- — dry lime partly absorbs the water content from the soil,
- — heat is generated,
- — carbonate production begins in the lime layer.

Because of chemical and thermal interactions involved, the increase in water content decreases temporarily the shear strength of the lime layers. On the other hand, since the CO_2 concentration in the air is higher in the vicinity of the soil surface than its average value (it can reach 1000 percent), rapid carbonate production ensues. The compound interactions increase the soil strength and a reinforced earth structure comes into existence.

In spite of ingenious solutions like the above, we think that the most common reinforced constructions will remain connected with mechanical interactions. Therefore, in the following we shall survey the most important factors influencing the behaviour of reinforcement.

3.3.1 Properties of reinforcing inclusions

The geometry (cross section, surface, shape, etc.) of reinforcement can be very diverse. Strips, grids, fabrics, webs, threads or fibres of very different materials (Fig. 3-4) can be arranged in a variety of ways to create reinforcing membranes. Engineering technology and industrial expertise must be coordinated to ensure the successful specification of constructional requirements and dimensioning considerations within the available technical resources.

Adhesion and surface friction occurring at the soil-inclusion interface play a paramount role in the interaction of soil and reinforcement. Therefore, among the geometrical characteristics, the surface must be emphasized, and the contact mechanisms should be investigated separately. The prerequisite of dimensioning is proper definition of the coefficients describing adhesion and frictional resistance; the required data can be determined by laboratory and site experiments.

Two experimental arrangements have become universal: in the pull-out test the inclusion is pulled out from a soil block kept at rest; in the shear experiment the soil block is moved on the surface of the fixed inclusion (Fig. 3-5). In principle, these experiments can be carried out on site and under laboratory conditions. However, because of technical difficulties, site shear experiments are seldom reported.

The mechanical properties of compound soil and inclusion blocks can best be described by their stress-strain relationships, shown in Fig. 3-6, where apparently great differences in magnitude and shape can be found. This diversity explains why difficulties are experienced when the mechanism of laminated reinforcing is investigated.

The stiffness of materials with different characteristics of expansibility is well correlated to their friction and adhesion. Therefore, characteristic values for their mechanical behaviour can be described by the range and magnitude of these quantities. In this section we focus our attention on this question. Many reinforcing materials exhibit time-dependent

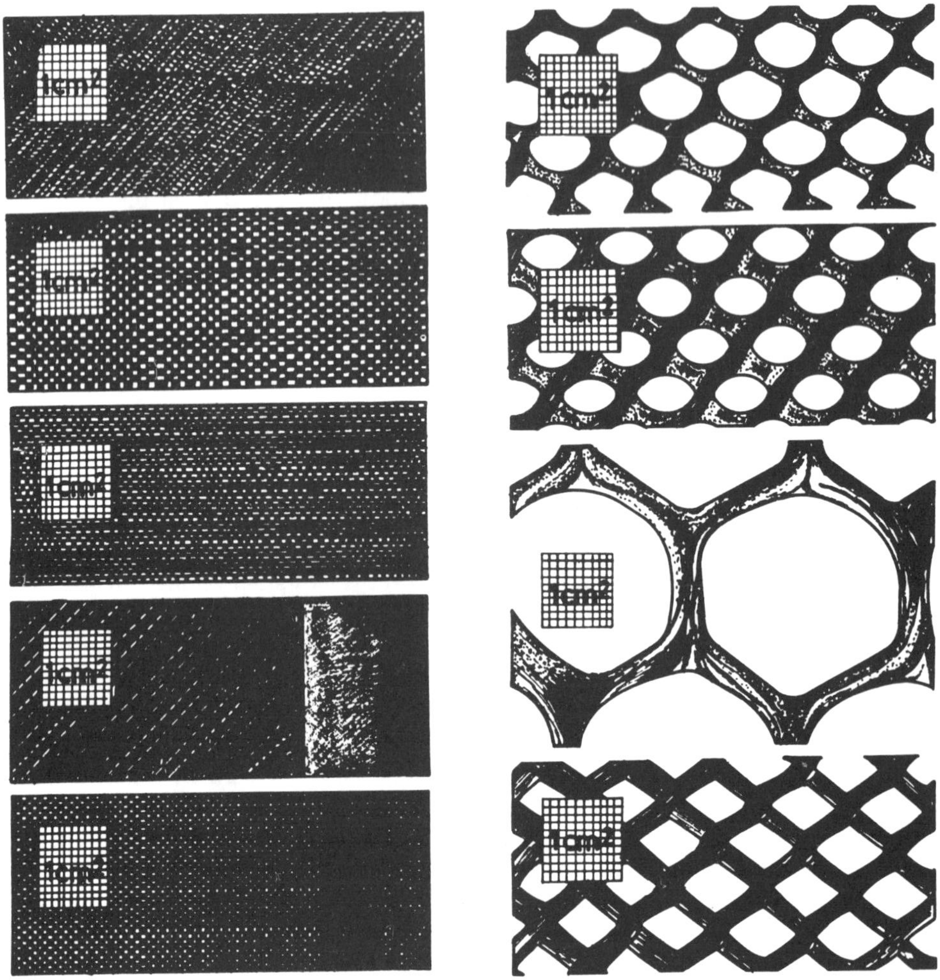

Figure 3–4 Soil reinforcing nets and grids (*Rankilor, 1981*)

properties which differ from those of soils. Consequently, we need to investigate also the factors that influence mechanical behaviour through time. We refer to the effects caused by heat, chemicals, sunshine, biological phenomena, etc. This field of research will not be dealt with in this book, but we note that a good survey of reinforcing materials is available (*Rankilor, 1981*).

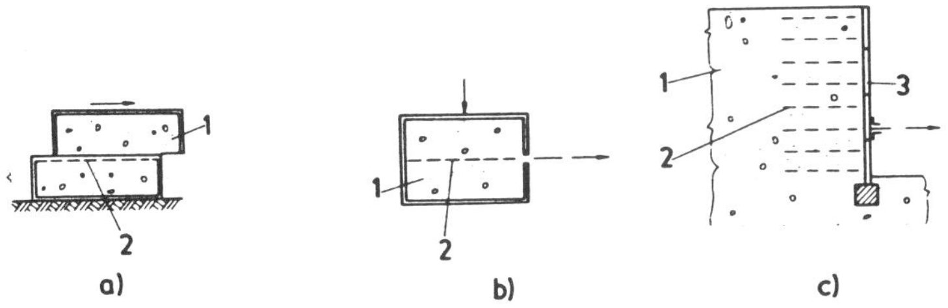

Figure 3–5 Experimental investigation of frictional resistance
a) laboratory shear test; *b)* laboratory pull-out test; *c)* site pull-out test;
1 soil; *2* reinforcement; *3* wall

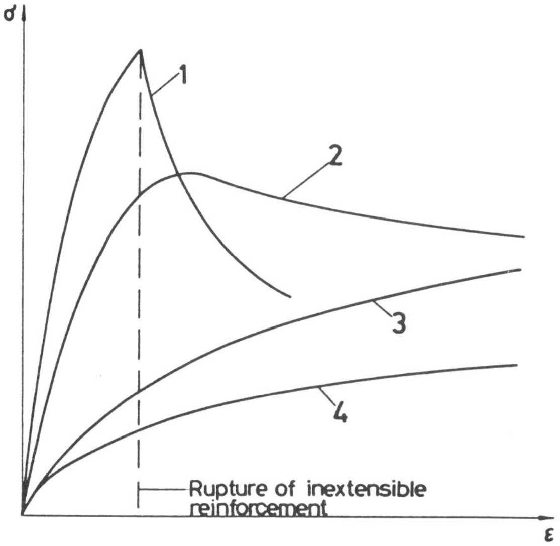

Figure 3–6 Characteristic stress–strain relationships
1 dense sand with stiff reinforcement; *2* dense sand with expansive reinforcement;
3 loose sand with expansive reinforcement; *4* loose sand without reinforcement;
σ tensile stress; ε strain

3.3.2 Straining capability of soil and reinforcement

Quality of interaction between soils and inclusions is greatly influenced by their relative stiffness:

— for strong reinforcements stiffer than soil by several orders of magnitude, surface effects can restrain the relative displacements along the interface to small strains only;

— for soft reinforcements with stiffnesses similar in magnitude to soils, the strains occurring in the soil and reinforcement may remain compatible during loading and only negligible relative displacements arise along the interface.

In most practical cases these two mechanisms appear together. The 'constitutive laws', (the stress–strain relationship) pertinent to the soil and inclusion materials are superimposed. Because the mechanical behaviour is controlled by the strains and also the type of interactions varies along inclusions, there can be different zones where different mechanisms act under the same working load. The extent of these zones may change with time.

The situation becomes more difficult where the stretching of the reinforcement is influenced by soil properties. Embedded and interlocked grains around fine meshes can strongly influence the strains, depending on the grain size distribution, effective diameter, etc. Plastic grids may creep, or relax, and soil consolidation may become bound up with viscoplastic behaviour of the reinforcement. Finally, but not least, these phenomena, as a rule, are irreversible and nonlinear and loading may be cyclic or dynamic.

In consequence we are unable to produce mechanical models of the same rigour as other civil engineering disciplines. Even some important fundamental questions have not yet been answered in this field. Therefore, the engineer must accept crude simplifications and design and construction methods reflect many assumptions which may not always be consistent with each other. Because of these (well-defined or tacit) approximations, we shall resist the temptation to construct sophisticated models to take into account repeated loading, rheologic or dynamic effects, etc. In what follows, we shall assume that the change of state of earth retaining walls can be considered as a slow monotonic process at low stress levels. Attention will be focused on three basic questions:

— the characteristic behaviour of expansive and stiff inclusions,

— the role of soil density,

— the features of basic inclusion types.

We feel that proper analysis of these factors makes it possible to find reliable solutions to the vast majority of engineering problems.

Among the investigations performed to describe the mechanical behaviour of earth structures strengthened with laminated reinforcement, the experiments reported by *McGown, Andrawes* and *Al-Hasani* (*1978*) are of great importance. In this investigation, inclusion response was described with respect to relative strain and initial soil porosity. Samples were examined in plane strain conditions (Fig. 3–7). 'Leighton Buzzard' sand was used with aluminium foil, aluminium grid and geotextile (type T140) as reinforcement. Load–deformation diagrams of the reinforcing materials are shown in Fig. 3–8. It can be seen that the strain characteristics of the metal reinforcement and geotextile differ significantly.

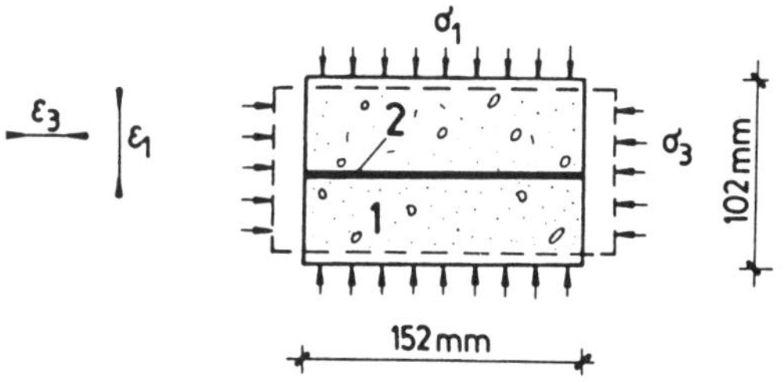

Figure 3–7 Test arrangement applied by *McGown* (*McGown, Andrawes, Al-Hasani, 1978*)

1 soil; *2* reinforcement

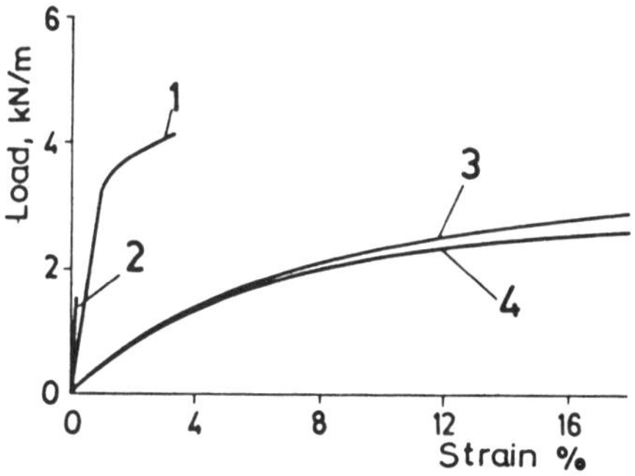

Figure 3–8 Stress–strain diagrams of reinforcing materials

1 aluminium grid; *2* aluminium foil; *3*, *4* geotextiles

(T140 Stabilenka: density 140 g/m^2, rupture strength 4800 N/m, rupture strain 105%)

In the experiments, the inclusions were placed horizontally in the first principal plane. The confining pressure (σ_3 = 70 kPa) was kept constant, while the vertical pressure was varied. Two glass plates on opposite sides of the sample were marked by square grids; vertical and lateral displacements were measured optically. The following aims were identified:

— the relationship between external loads and internal strains,
— the analysis of internal strain fields,
— the role of the reinforcing material's extensibility (eventually for classification purposes).

The porosity of the soil was varied so that samples with different initial void ratios were investigated. Figures 3–9 and 3–10 show the relationship between vertical strain ε_1 and external load characterized by the stress ratio σ_1/σ_3.

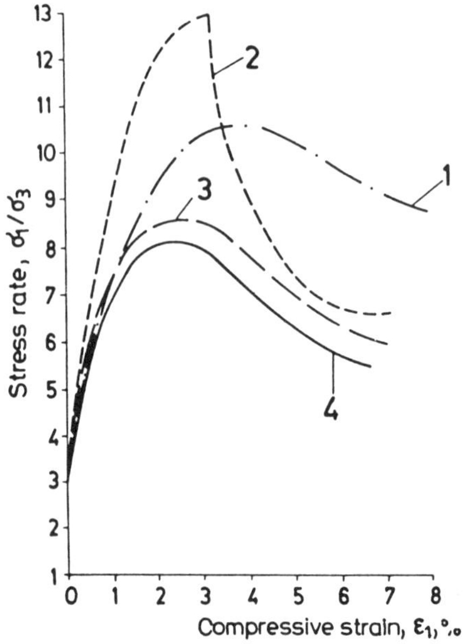

Figure 3–9 Stress–strain curves for dense sand (*McGown, Andrawes, Al-Hasani, 1978*)
1 sand and geotextile; *2* sand and grid; *3* sand and foil; *4* pure sand

For dense soils (Fig. 3–9), the usual hardening process occurs in both the original soil and the reinforced soil, with a more or less steep initial response. A characteristic strain of 2...4% can be identified with the maximum load. It can be stated that

— reinforcement increases the sample's strength,
— metal inclusions influence the characteristic strain only slightly, while geotextiles yield significantly different responses,
— rupture of the aluminium grid causes instant and substantial loss of strength.

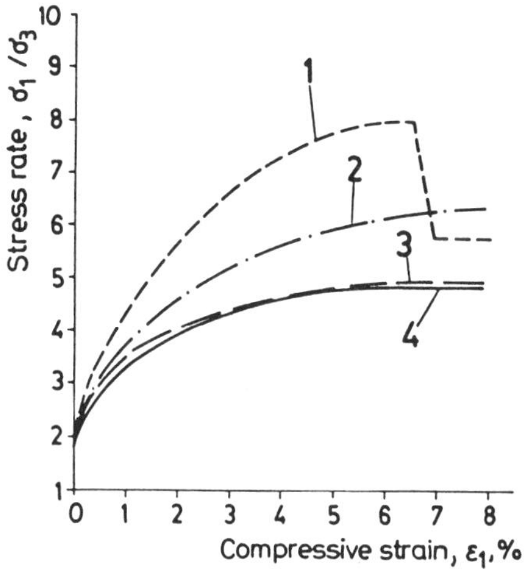

Figure 3–10 Stress-strain curves for loose sand (*McGown, Andrawes, Al-Hasani, 1978*)
1 sand and grid; *2* sand and geotextile; *3* sand and foil; *4* pure sand

For loose soil samples, hardening develops smoothly (Fig. 3–10). The strengthening effect of the reinforcement and sudden loss in strength after rupture was found in these cases, as well. For strains greater than 6...8% the curves became almost horizontal. Further characteristics of the mechanical behaviour can be demonstrated using the test results for samples with different initial densities. The plot of maximum loading ratio σ_1/σ_3 versus initial porosity is shown in Fig. 3–11. It is worth noting that the curves representing dense sand, reinforced or not, are similar but the reinforcing effect is more explicit for loose soils, even in the case of geotextile reinforcement. Of course, strains are not considered here.

The internal deformations which develop in the interior of the samples can be derived from the displacements measured at the grid points on the sides. The dense initial state is worthy of consideration because two significantly different stages can be distinguished subsequently. In the first stage, up to the peak value of the stress ratio, hardening occurs, which is followed by softening in the second stage. The results are compared with the reference behaviour of a dense soil without reinforcement.

Displacement vectors measured in the first and second stages in the case of a dense sand (before and after the peak stress ratio is reached) are shown in Fig. 3–12*a*. The displacement field is symmetric before reaching the peak value and no distinct discontinuity occurs. In the second stage, two shear zones can be observed and the displacement distribution indicates the motion of a rigid body. For the soil sample

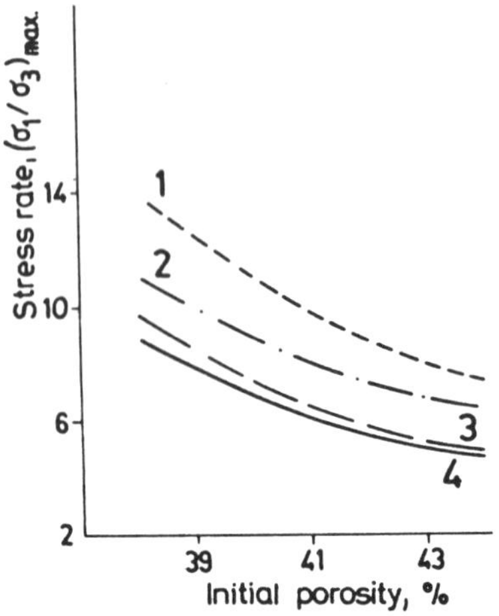

Figure 3–11 Relationship between maximum loading ratio and porosity (*McGown, Andrawes, Al-Hasani, 1978*)

1 sand and grid; *2* sand and geotextile; *3* sand and foil; *4* pure sand

reinforced with aluminium foil, similar displacement distributions occur, with more definite differences between the 'before peak' and 'after peak' stages (Fig. 3–12*b*).

Significant differences may be observed in the case of the samples reinforced with geotextiles (Fig. 3–12*c*). There, symmetry remains and no characteristic, singular surfaces can be observed in the displacement field.

Distortion (shear) strains determined from the displacement vector fields are shown in Fig. 3–13. The anticipated symmetry before the peak stress ratio is reached can only be guessed at in the case of unreinforced dense soils (Fig. 3–13*a*). After the peak value is reached, characteristic shear zones appear in accordance with the given boundary conditions. In the case of aluminium foil reinforcement, the distribution pattern before the peak turns out to be different: high shear stresses develop around the inclusions. Similar, even more significant concentrations can be seen in the 'after peak' state of the shear strain distribution (Fig. 3–13*b*). Another strain pattern was found for samples with geotextile reinforcement (Fig. 3–13*c*). Before the peak value of the stress ratio is reached, shear strain concentrations can be observed around the inclusions. However, the 'after peak' distribution is rather obscure, the anticipated sliding surface can only be assumed.

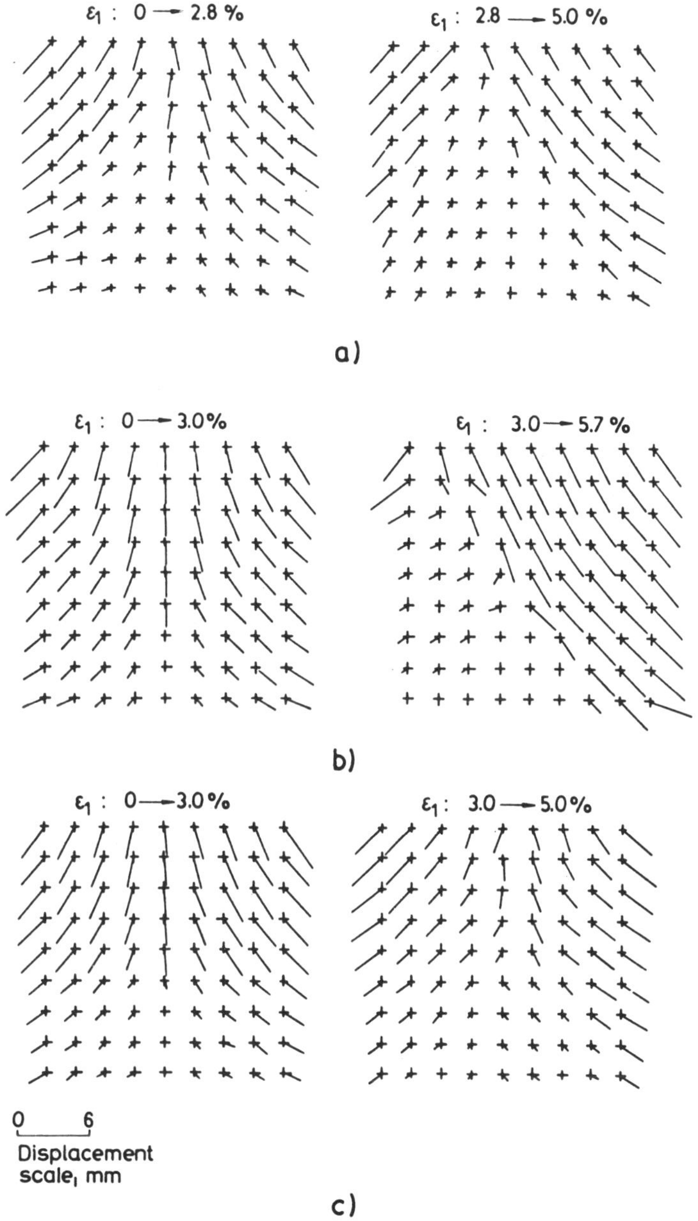

Figure 3–12 Displacement fields in dense sand (*McGown, Andrawes, Al-Hasani, 1978*)

a) pure soil; *b)* soil reinforced with aluminium grid; *c)* soil reinforced with geotextile

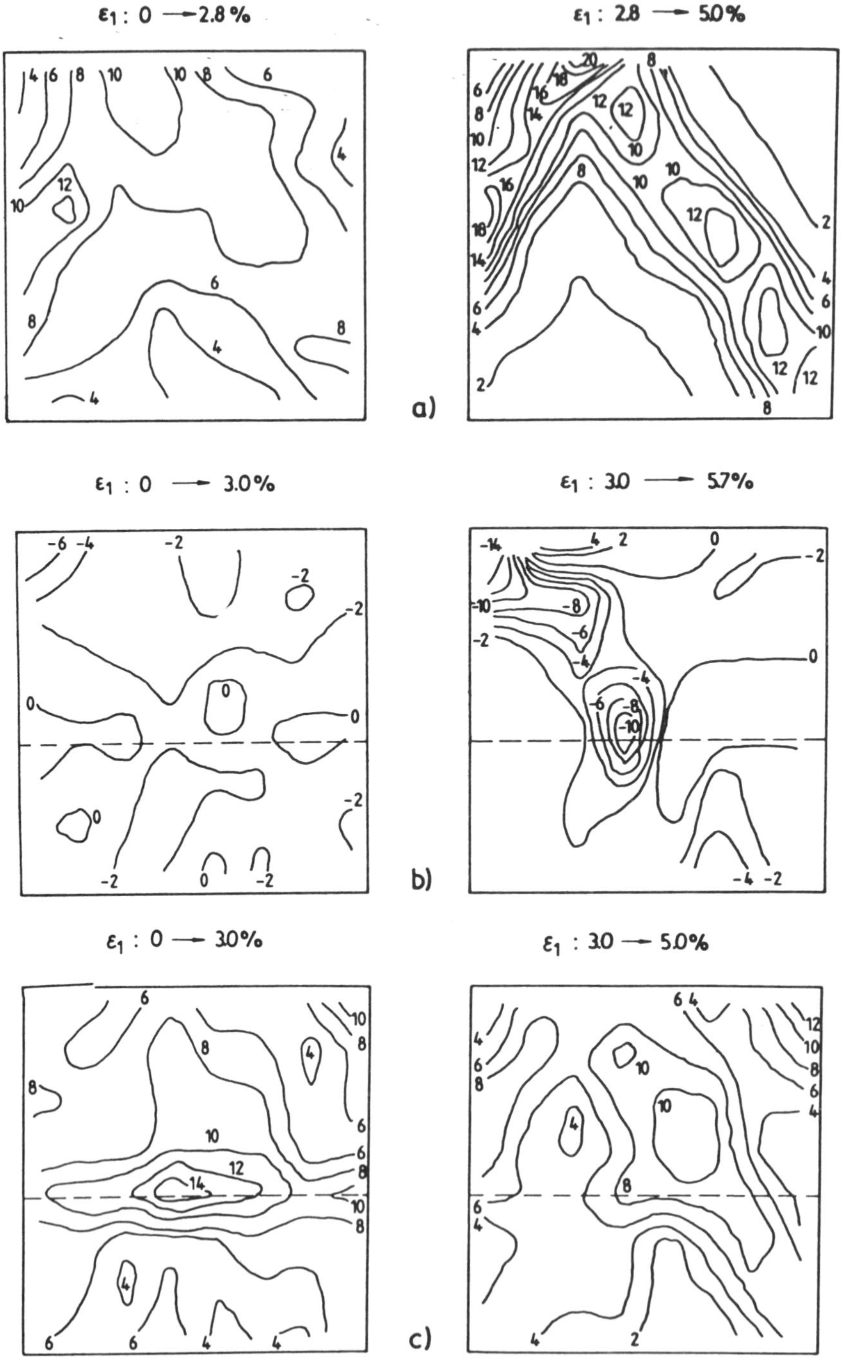

Figure 3–13 Maximum shear stress distribution in dense sand (*McGown, Andrawes, Al-Hasani, 1978*)

a) pure soil; *b)* soil reinforced with aluminium grid; *c)* soil reinforced with geotextile

Differences can be observed in the volume change distribution, too. For dense sand, some loosening appears before the peak stress ratio is reached. After peak—in the sheared zones—definite loosening occurs (Fig. 3–14*a*). Aluminium reinforcement initiates loosening before peak; in the volume strain zone after peak (especially in the case of inclusion rupture) an extensive loosening zone can be observed (Fig. 3–14*b*) which coincides with the sheared zone. Another pattern has been found for dense soil reinforced with geotextiles (Fig. 3–14*c*). Before reaching the peak value of the stress ratio, some volume decrease was observed around the reinforcement (partly due to self-compression of the inclusion). After peak, the distribution becomes similar to that of sand without reinforcement, but the strains are smaller than those measured in the reinforced sample.

The conclusions can be summarized as follows:

— the inclusions increase the strength of samples depending on soil density and elasticity of reinforcement,
— the magnitude of the expansion, which can vary widely, is reflected in very different reinforcing mechanisms,
— the less expansive inclusions increase to a greater extent the soil strength in their surroundings before rupture. Following rupture, their reinforcing effect almost disappears. These inclusions significantly constrain the strains before rupture; after it, they lose their influence. The kinematics of rupture is very similar to that of the reference (unreinforced) material,
— highly expansive inclusions are effective in a wide range of conditions, up to large strains. Rupture, as a rule, is not connected with development of rigid soil regions and shear surfaces.

The effectiveness of reinforcement decreases if it is not parallel to the direction of principal strain. Any reduction in effectiveness can be estimated as a function of the angle between the direction of the reinforcement and the principal tensile strain. If it exceeds 40...50 degrees, a weakening effect of 'reinforcement' has to be taken into consideration (Fig. 3–15). Reinforcement lying in the shear zones can be very disadvantageous, if the resistance of the soil-reinforcement to shear is lower than the natural internal shear resistance of the soil. Movement of a rigid, shifting soil block can occur in these cases at a relatively low stress level due to the presence of 'reinforcement', whilst obviously the bearing capacity of the structure decreases.

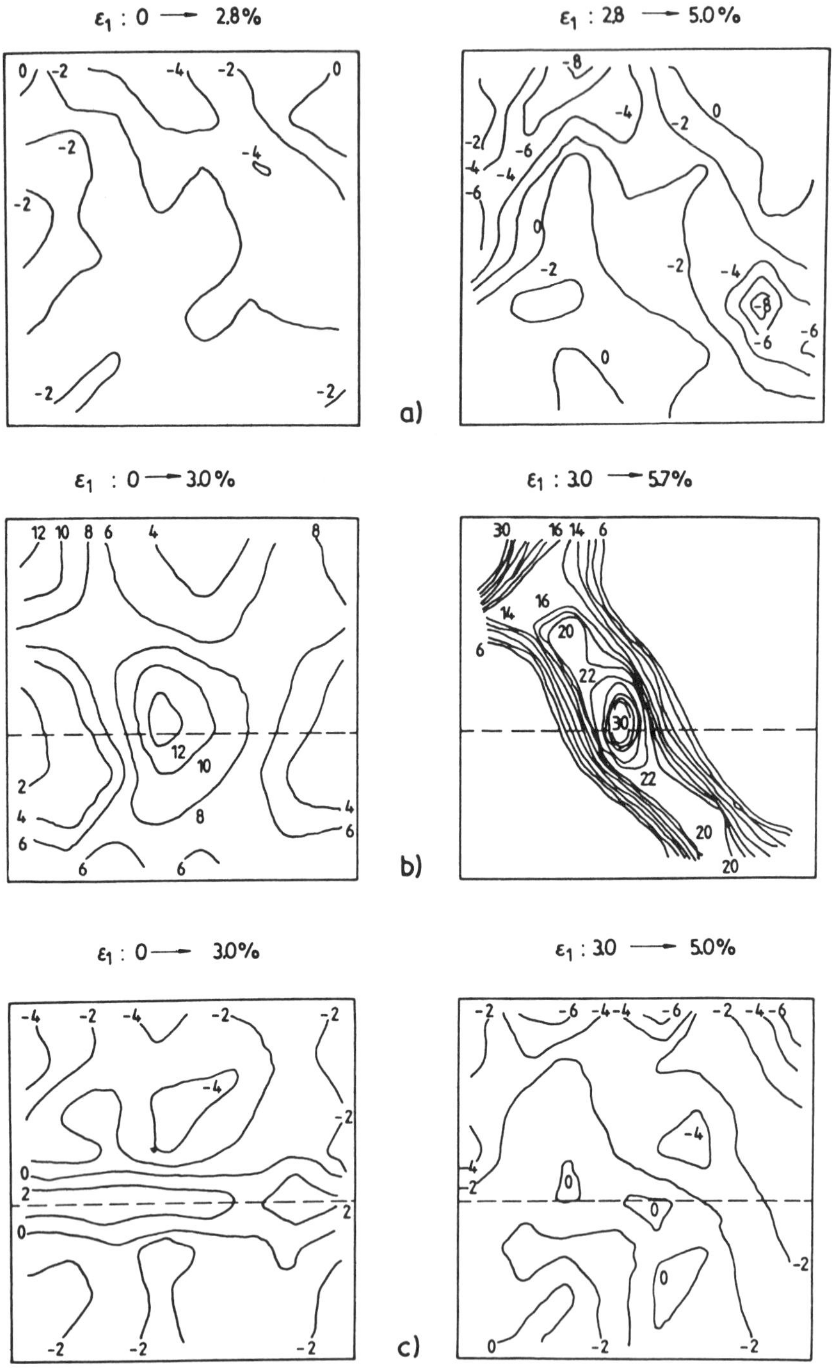

Figure 3–14 Volume change distribution in dense sand (*McGown, Andrawes, Al-Hasani, 1978*)

a) pure soil; *b)* soil reinforced with aluminium grid; *c)* soil reinforced with geotextile

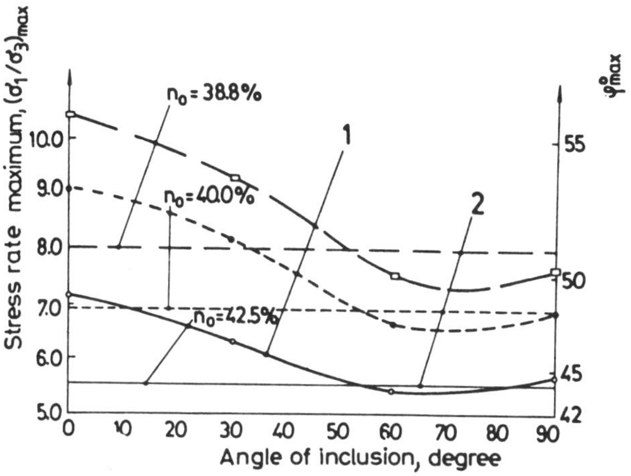

Figure 3–15 Reinforcing effectiveness changing with the reinforcement inclination angle (*McGown, Andrawes, Al-Hasani, 1978*)

1 sand and geotextile; *2* pure sand; φ_{max} maximum friction angle; n_0 initial porosity

3.3.3 Materials for reinforcing layers

The variety of materials appropriate for layered reinforcing methods is very rich, since almost any material of some tensile strength can improve the mechanical properties of the soil. Even a material whose compressive strength is greater than that of a given soil may bring about a reinforcing effect.

Reinforcement materials can be classified as follows:

— metals, like alloyed or non-alloyed steels, aluminium,
— products of (organic or inorganic) plastics,
— layers saturated with bonding materials (like cement, chemicals, bitumen),
— reinforcements of vegetable origin (even alive).

Metals are the most common and most precisely defined engineering materials, and their properties are well known and described in various monographs (*Alexander, Brewer, Rowe, 1987*). Several questions, connected with soil–reinforcement interaction, will be discussed in detail in Chapters 5 and 6.

More detailed information is needed for plastics. In comparison with traditional construction materials their modulus of elasticity can be less by two or three orders of magnitude and their performance may strongly depend on loading conditions and the passage of time. Therefore, the common assumption of linear elasticity (often tacit in geotechnical design) may not be valid when plastic reinforcements are used. Additionally, nonlinear stress–strain relationships and creep effects should be taken into account.

In accordance with our general purpose, we will not dwell on the properties of materials used for grouting or stabilization. The interested reader may find detailed descriptions in monographs focusing on the technology of these reinforcing methods (*Kézdi, 1967*). The variety of (and market for) grouting materials widens continuously, but their influence on mechanical behaviour and construction is minor.

The inventiveness of designers and contractors often results in very interesting retaining walls reinforced with materials of vegetative origin (such as roots, leaves, stems, branches). We have described these types of reinforcements in Section 2.2. Although their use is impractical for industrialized construction technology, they have a growing role—both for economic reasons and for their environmental advantages—their increasing use can be anticipated in the future.

3.3.3.1 Plastic geotextiles

The variety of threads produced by the chemical and processing industry offers rich possibilities for using them in engineering constructions. Several sorts of fabric are classified in Table 3–1, but the classification cannot reflect the output of 25...30 major chemical enterprises and countless smaller factories, which bring several hundreds of products to the market. Most of the raw materials are common polymers produced for other applications (like polyamide, polyester, polypropylene, PVC, etc.).

Table 3–1 Types of plastic geotextiles

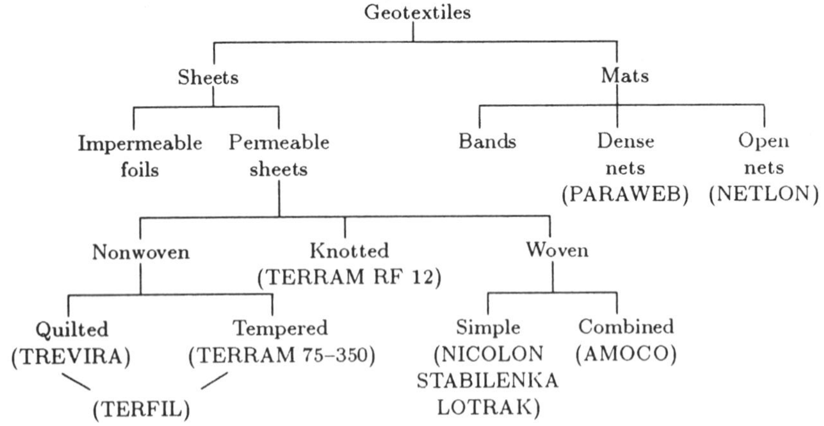

Of course, the cost of the raw materials used in construction must be paid by the user. The mechanical properties of geotextiles (even of those offered at the same price level), however, may differ significantly from each other. Therefore, the designer must have good knowledge of design practice and the mechanical properties of reinforcement. One has to avoid at least two errors:

— over-reinforcement (either in strength or quantity) may result in uneconomic construction,

— the reinforcement may degrade with time because of unknown unfavourable properties.

With respect to the mechanical behaviour of structures, the interaction between soil and reinforcement at their interfaces has paramount importance. The reinforcement counterbalances the external loads by development of adhesion and frictional resistance on its surface. Its bearing capacity can be exploited only if the roughness of the surface attains some threshold value. This requirement is generally known and manufacturers take it into account when the geometry of the reinforcement's cross section is fashioned. Of course, there is no surface geometry absolutely suited to all possible sorts of soils (and for all possible soil conditions) but this is not a necessity.

The situation is less satisfactory with respect to the stress–strain parameters used for design. Most products are classified according to the standards of the textile industry (*Postle, Carnaby, de Jong, 1987*). For example, the tensile strength of a plastic strip used in the building industry is measured in a specimen of 50 mm width, a test which originates in the textile industry. While various tests conducted on these materials in isolation provide a general picture for comparing various products, it is hardly ever possible to make a reliable estimate of the tensile modulus of a material embedded in soil. Consequently, tests to describe the interaction of soil and reinforcement have been invented and are used both in laboratory and (rarely) under site conditions.

The simple uniaxial tensile experiment is not adequate to determine a stress–strain relationship for reinforcing grids, mats or textiles, since the contraction of the sample in the transverse direction may strongly influence its elongation and so low tensile moduli will result. Experiments show (*van Leeuwen, 1977*) that in the case of non-woven polyamide and polyester webs, the tensile strengths measured for samples of constrained cross contraction, as exposed in practical applications, can be two or three times higher than those observed for unconstrained samples in the laboratory (Fig. 3–16). For samples whose elongation strongly depends on the direction of loading, contraction can be prevented in the transverse direction by using special techniques. It is generally accepted that samples of minimum 200 mm width should be used for these experiments.

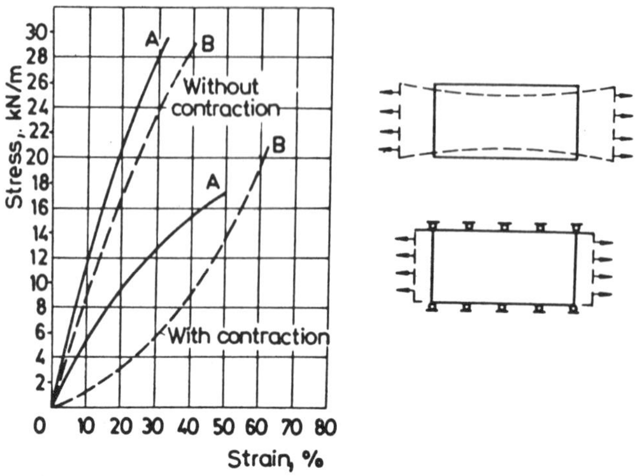

Figure 3–16 Stress-strain diagrams of different geotextiles (A, B) influenced by test conditions (*van Leeuwen, 1977*)

Other sophisticated (and arbitrary) experiments for tearing, twisting, crushing, etc. customary in the textile industry are less important from our point of view. Results of these experiments can be taken into account where local effects or forces like concentrated loads, coarse grains, etc. threaten the integrity of reinforcement in the soil and their possible consequences are to be assessed (*Sissons, 1977*).

Experimental data determined for non-woven geotextiles (after *Alfheim* and *Sorlie, 1977*) are given in Table 3–2. These data give a general impression of the magnitudes of the strength parameters. Pertinent values for more recent reinforcing grid products are usually available in advertisements published by manufacturers keen to promote them for engineering purposes (Table 3–3).

Experiments relating to grids embedded in soils can be classified into the following groups:

— triaxial tests for determining the required number, distance, reinforcing efficiency of sheets; the results of these tests is a modified modulus of compressibility, usually greater than that of the soil, reflecting the reinforcing arrangement,

— direct shear tests to determine adhesion and frictional resistance parameters relating to the soil–reinforcement interface; the coefficient of friction can be determined in this way,

— pull-out experiments in the laboratory or at site to find the force required to cause relative displacement of reinforcement with respect

Table 3–2 Characteristic parameters of non-woven geotextiles

Name	Type	Product parameters		Laboratory characteristics*		
		1	2	3	4	5
COLBOND (D, NL)	P 250 P 450	250 450	14/8 27/16	9.8 19.4	12 27.5	40 42
FIBERTEX (DK)	S 170 S 300	170 300	6.4/8 12/16	4.8 9.2	10.9 21.8	65 70
TERRAM (GB)	PRF 70 PRF 140 PRF 280	70 140 280	2.6 4.9 9.4	— 3.2 6.9	— 5.5 10.6	— 70 50
TREVIRA (D)	200 300 400 500	200 300 400 500	10 15 24 27	4.0 7.7 13.6 19.2	15.3 23.2 32.7 44.5	73 78 82 73
BIDIM (F)	U 14 U 44	150 340	10 26	7.4 12.6	13.6 32.1	53 76

Legend: * — Norwegian Road Research Laboratory,
1 — specific amount of material, g/cm^2,
2 — rupture strength (in longitudinal and cross direction), kN/m,
3 — tensile strength at 20% strain, kN/m,
4 — maximum tensile strength, kN/m,
5 — rupture strain.

Table 3–3 Characteristic parameters of geotextiles (production data)

Name	Type	Specific weight g/m^2	Rupture strength kN/m	Load at 10% strain kN/m	Strain at % of rupture strength	
					50%	100%
TERFIL	I. II. III.	250 400 500	6.5/5.0 14.0/10.0 18.0/14.0	— — —	— — —	40...70/ /70...110 50...80/ /70...110 60...90/ /70...110
NETLON	H–08 H–11 H–12	540 730 600	8/— 10/— 9/—	5.5 6.5 6.0	— 2.0 3.5	100 20 20

to the surrounding soil; a generalized value for the coefficient of friction can be found by this experiment (used in internal stability investigations).

Triaxial experiments are performed, as a rule, on samples of large diameter. The number of reinforced layers, soil density and confining pressure are the usual variables. The role of confining pressure is of great importance and must correspond to the 'at rest' value anticipated at site in order to avoid using excessive values for the internal friction angle in the reinforced soil. Typical behaviour of soil samples reinforced with grids is illustrated in Fig. 3–17 (*Broms, 1977*). The soil–reinforcement interaction can be characterized as described by *McGown et al* (see Fig. 3–13 to 15).

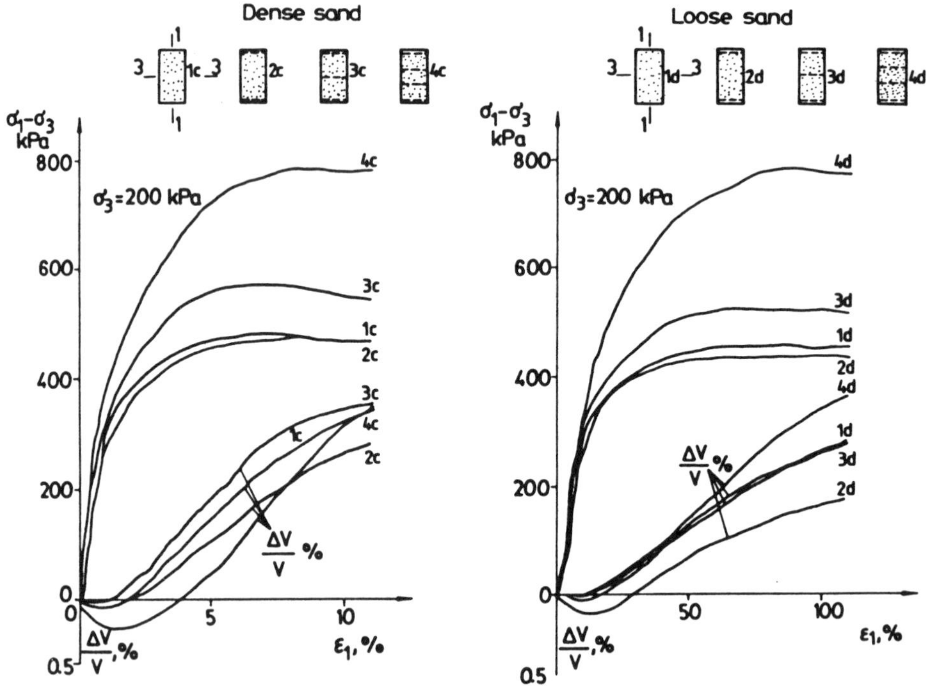

Figure 3–17 Triaxial stress–strain results (*Broms, 1977*)
a) dense sand; *b)* loose sand

Results of both direct shear and pull-out tests in the laboratory are appropriate for determining frictional parameters, which can then be used in limit state theories. However, the practical application of these parameters can be disputed, because the boundary conditions of the laboratory tests differ substantially from those met in practice. The importance of this factor can be illustrated by the multitude of different experimental arrangements invented by researchers (Fig. 3–18; *Hausmann, Ring, 1980*). Evaluation of the experimental results is simple in principle: the ratio of shear to normal forces can be taken as the tangent of the 'apparent' angle of friction. However, one faces several difficulties:

— in the case of direct shear experiments the effect of volume change induced by shear may exceed the effect of any reinforcement,
— the distribution of frictional resistance along the inclusion's surface may be variable, the pull-out resistance may be influenced significantly by the shape of the inclusion (e.g. its waviness) and by the boundary conditions which constrain the behaviour of the surrounding soil,
— in the case of several reinforcing materials (e.g. grids or mats) the concept of surface resistance itself may be questioned because of interlocking effects,
— other non-mechanical effects (like moisture) can influence the value of measured resistance.

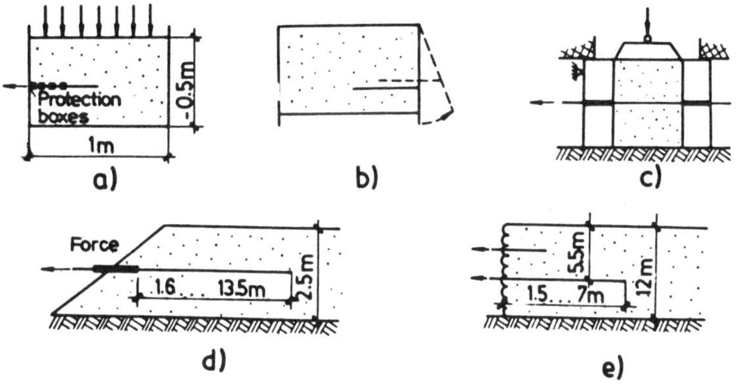

Figure 3–18 Characteristic pull-out test arrangements (*Hausmann, Ring, 1980*)
a) pull-out box; *b)* moving wall pull-out; *c)* modified direct shear test;
d) large scale site pull-out test; *e)* pull-out of 'blind' inclusions

The importance of boundary conditions has been demonstrated by X-ray investigations (*Schwab, Pregl, Broms, 1977*). Controversial observations have been reported with respect to the influence of friction. In some experiments performed in France, the apparent angle of friction determined by direct shear test appear to be underestimated. Pull-out tests, carried out in the USA, have shown the opposite and resulted in low frictional resistance. Several authors state that the angle of friction on the reinforcement–soil interface cannot be greater than the internal angle of friction of the soil. Any greater experimental result than that must be due to other side-effects (*Schlosser, Juran, 1981*).

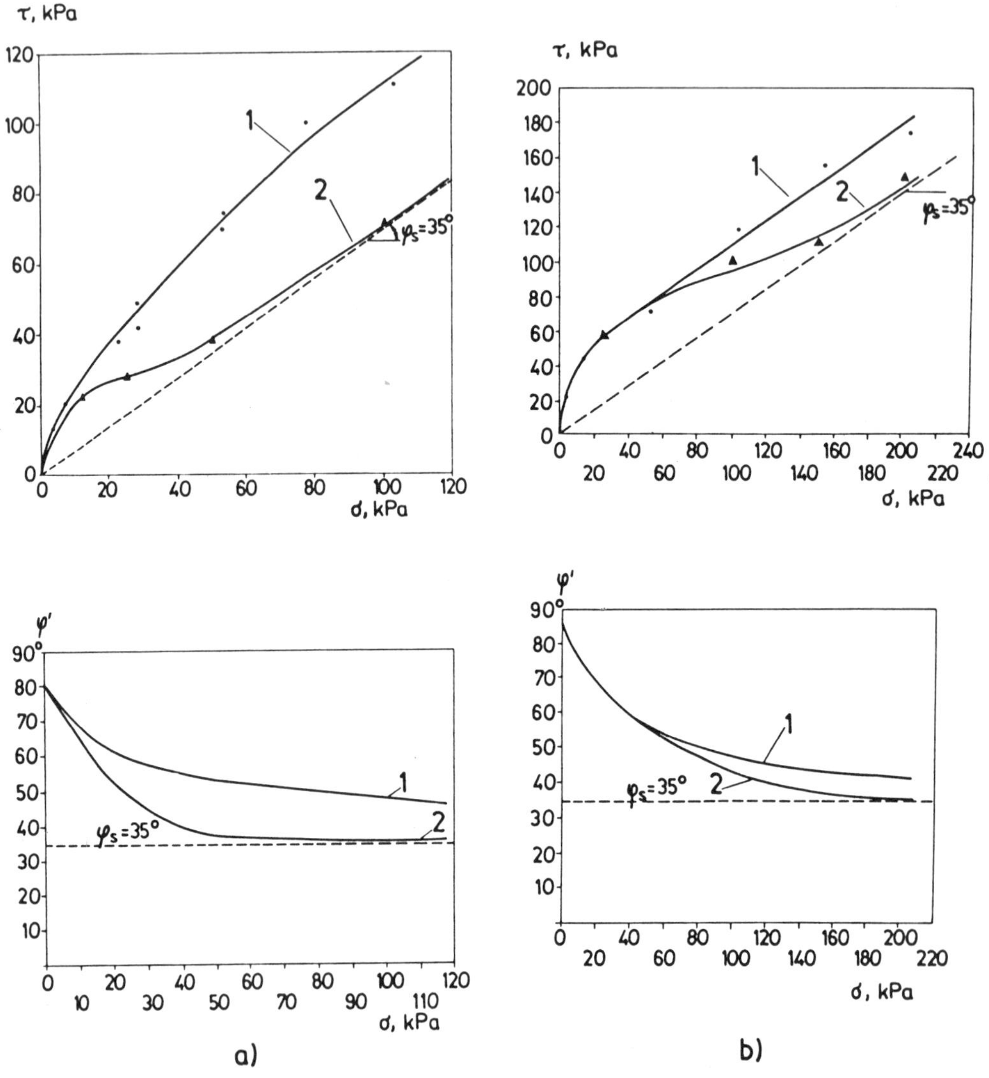

Figure 3–19 Test results related to experimental arrangements (*Ingold, Templeman, 1979*)

a) steel dusted with sand; *b)* polyolefin grid; *1* pull-out test; *2* shear test; φ_s internal friction angle of sand; τ shear stress; σ normal stress; φ' apparent friction angle

The differences between experimental results were investigated for several reinforcing materials and using various experimental arrangements by *Ingold* and *Templeman* (*1979*). Curves shown in Fig. 3–19 illustrate the importance of the experimental arrangement which remains predominant and almost independent of the characteristics of the reinforcement. We are inclined to conclude that no soil–reinforcement interaction can be described properly with a single frictional parameter, no matter how carefully the tests are conducted.

Further research is required to explore all the kinematical phenomena occurring in the surroundings of contact surfaces. This work is also required for the efficient application of powerful computational techniques. In using finite element methods, for example, special interface elements could be derived using these data. Until these questions are answered, the pragmatic opinions like, "the most advantageous experiment is the pull-out test because it reflects all the sophisticated effects unmeasurable properly in other ways" should not be rejected.

3.3.3.2 Creep parameters for plastic grids

In analyses of the static and kinematic behaviour of plastic products used for soil reinforcement, creep and relaxation effects must not be neglected. The quantity of required reinforcement depends mainly on permanent loads and acceptable design deformations. Therefore, information about parameters connected with long-term strains and stress relaxation is essential. However, at the time being, neither experimental results nor relevant production data are available, although design methods are well advanced and appear to give satisfactory solutions. Consequently, in this section only the fundamental effects and some of the most significant experimental results will be discussed.

The importance of strains in plastics has been emphasized. In fact, the mechanical behaviour of plastics can be better described by focusing on strains instead of stresses. From experiments performed on elementary threads, the strain ε_t varies with time t and can be approximated by the linear expression:

$$\varepsilon_t = \varepsilon_o + b \log t$$

were ε_o stands for the initial (instant) strain, and b denotes the creeping coefficient; (t is measured in minutes; *Finnigan, 1977*). The creep coefficient depends on the level of loading (defined as the ratio of actual load to the load causing rupture). Experimental data referring to polyester and polyamide grids are given in Fig. 3–20. It can be seen that for loads not higher than half of the rupture load the creep parameter can be taken as $b = 2.0$. Hence, the initial strain increases one percent in 10^5 minutes, and several decades must pass before the strains reach a value causing rupture.

In engineering practice, working loads usually do not exceed the half of rupture values (coefficients of safety against failure as a rule are greater than 2.0). This consideration, therefore, shows that

— for provisional constructions no danger can be anticipated at loads not greater than 50% of the rupture load,
— during design for example for a design life of 50 years one should lower the loading level or use materials with smaller creep coefficients to achieve a sufficient degree in safety.

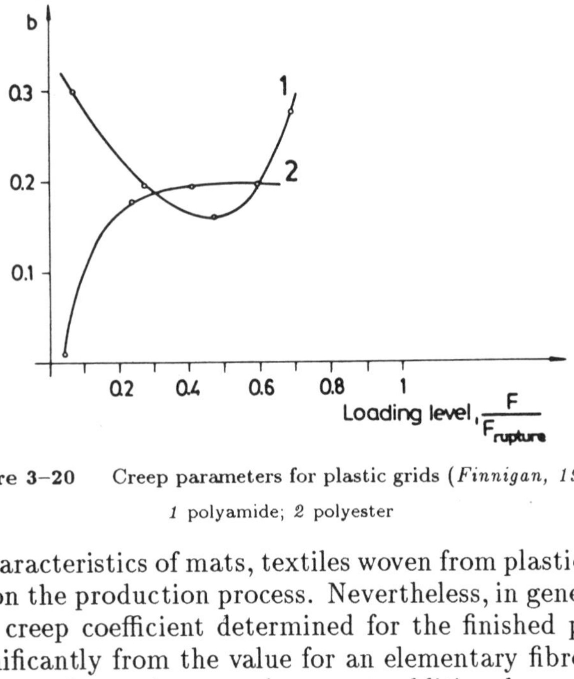

Figure 3–20 Creep parameters for plastic grids (*Finnigan, 1977*)

1 polyamide; *2* polyester

Creep characteristics of mats, textiles woven from plastic fibres naturally depend on the production process. Nevertheless, in general the magnitude of the creep coefficient determined for the finished product does not differ significantly from the value for an elementary fibre. Therefore, we may estimate that only several percent additional creep strain will occur if the working load does not exceed one third of the rupture load.

Similar results were found for non-woven geotextiles, (*van Leeuwen, 1977*). The publication referred to earlier (with respect to the one- and two-dimensional tensile experiments) also reports on several long-term tensile investigations. Polyamide and polyester geotextiles loaded with forces as high as 60% of the rupture load did not lose their bearing capacity for several years. The instantaneous strain ε_o was about 80% of the total strain measured after some years.

3.4 LINEAR SOIL REINFORCING

In addition to randomly placed reinforcing elements distributed in three dimensions and two-dimensional reinforcing webs, linear inclusions can also be used for earth retaining wall constructions. Their basic characteristics are their rod-like geometry and linear reinforcing effect. The practical advantages of linear reinforcement arise in construction practice. In situations, where the soil should be kept in its natural state, the simplest solution is to place rods into bored holes. Simple modifications of common anchoring practice can be used therefore, to produce bar inclusions which interact with the soil along their full length. A real advantage of the rod's geometry lies in its decreased corrosion susceptibility: the surface to volume ratio is a minimum in comparison with any other shape. On the other hand, this cross section (mostly circular, rarely polygonal) is not very efficient; increased contact surface area and proper direction would improve the interaction between soil and reinforcement. Careful engineering analysis is required to find a proper balance between these contradicting aspects.

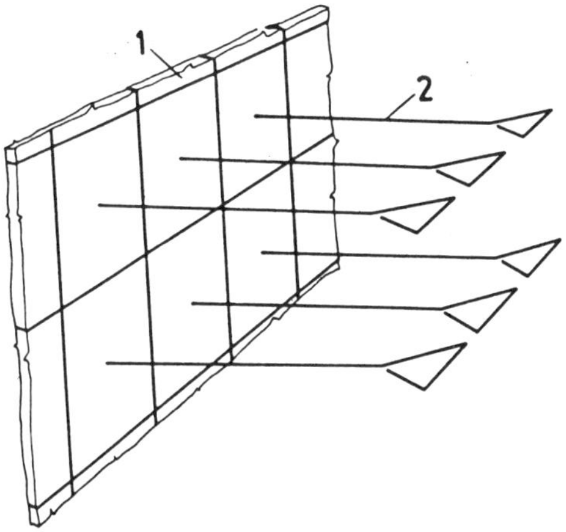

Figure 3–21 Wall reinforced with rods bent into special shapes
1 facing; *2* reinforcing–anchoring rod

Strip-reinforced soil retaining walls, as invented by *H. Vidal* may be mentioned here as examples, albeit the overall arrangement of the strips results in a two-dimensional structure. These types of constructions will be dealt with in detail in Chapter 5. Another interesting example can be seen in Fig. 3–21. Rods bent into special shapes were used to fix facing elements; both surface resistance along the full length of the rods and the anchoring effect of the triangular butt were taken into account (*Murray, 1981*).

In the case of cut-in walls the advantages connected with technology increase. Therefore, we devote a full subsection (6.1) to the widely used 'nailed soil' wall structure.

4

EARTH WALLS

Having arrived at the present Chapter, the Reader hardly requires further definitions of earth retaining walls. However, to prepare the way for the forthcoming chapters on the questions of structural arrangements, design and construction we still have to classify into two fundamental classes, namely, retaining and revetment walls. The differences will be analysed in detail in Chapters 5 and 6.

4.1 GEOMETRIC ARRANGEMENT OF EARTH WALLS

Where establishment of horizontal surfaces is desired on slanting ground, two alternatives are possible, as shown in Fig. 4–1.

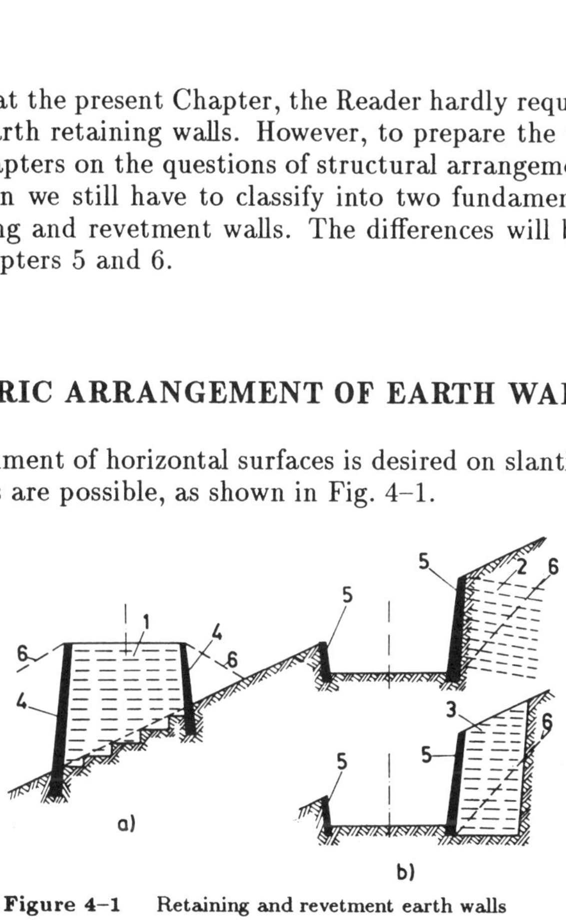

Figure 4–1 Retaining and revetment earth walls

a) retaining walls; *b)* revetment walls; *1* reinforced soil; *2* direct reinforcement; *3* reinforced backfill; *4* retaining wall; *5* revetment wall; *6* optional slope

In the case of embankments, instead of slopes, retaining walls may be required. The designer may find it appropriate to consider the wall and the supported soil mass as a whole and a combined structure can be designed and built (Fig. 4–1*a*). The natural environment of the embankment will, in general, remain undisturbed, and can be treated as a 'boundary condition'. Simple technologies are available for placing reinforcements and facing the earth mass.

Revetment walls are built (Fig. 4–1*b*) if circumstances require earth removal. Two alternative arrangements can be seen in Fig. 4–1*b*:

— earth removal and reinforced backfill (insert bottom),

— direct reinforcement of the undisturbed soil (insert on top).

The first alternative exploits the advantages of embankment construction technologies. However, the amount of earth removal may well exceed the minimum volume required by the geometry. On the other hand, technologies used for reinforcing earth masses are more complicated in general, and generating appropriate soil–reinforcement interaction may necessitate more sophisticated techniques.

4.2 STRUCTURAL VARIANTS OF EARTH WALLS

Construction of the various geometric alternatives described in Section 4.1. involves (at least to some extent) a number of different technologies. In the case of retaining wall stabilization, grouting, mechanical reinforcement (e.g. compaction) and the use of reinforcing inclusions are plausible solutions. Laminated arrangements (the use of webs, grids, strips and possibly concrete elements) may be practical for the latter alternative.

Where revetment walls are built, it is undesirable to disturb the natural state of the soil. Therefore, beside soil grouting, reinforcement techniques using inclusions, i.e. nailing or trenching can be recommended. Concrete element systems combined with earth-made structures may be economic as well.

4.3 STATICS OF EARTH WALLS

The mechanical behaviour of earth walls is rather complicated. Consequently, relevant static and dynamic models, design and computational approaches incorporate many assumptions. The degree of simplification depends on the experimental evidence and may be influenced by the expertise and computational skill of the designer.

First of all, a proper description of the behaviour of the designed structure is required. Most design methods rely on limit equilibrium methods; only occasionally are strains and displacements determined. This deficiency is, however, by no means a peculiarity in the design of earth walls but is typical of geotechnical problems in general, due to the complexity of the constituent material, namely the soil.

Among the different constructional arrangements, there are many solutions based on intuitive engineering considerations. Very often, the bearing capacity of these constructions is crudely estimated and the actual factor of safety remains undetermined and hidden. Nevertheless, in engineering practice, it is customary to assume high safety factors for earth walls. On the other hand, because of economic considerations, widespread research is being conducted to reduce the unavoidable overestimations by elaborating better static models.

Investigations connected with design can be classified into the two groups described in the following two subsections.

4.3.1 Investigation of external stability

As a first approximation, an earth-made retaining structure can be considered to be a gravity wall. Therefore, investigations into the motion of a rigid body are performed for all appropriate degrees of freedom (Fig. 4–2). Essentially, in these investigations, the analysis involves calculation of all forces and moments which cause the assumed movement and those which are acting against the motion. The ratio of the latter to the former defines the safety factor.

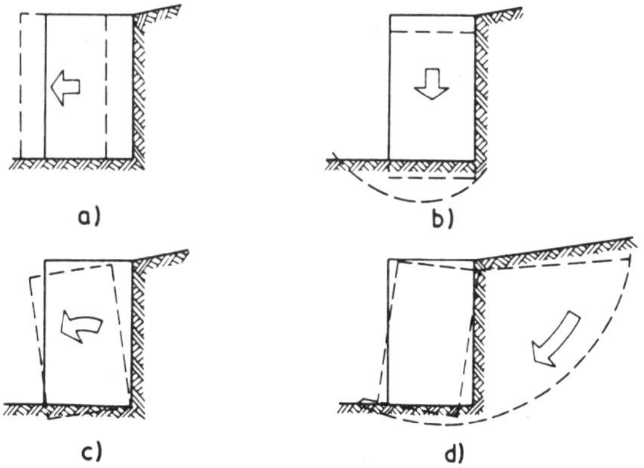

Figure 4–2 Kinematic modes for external stability investigation

a) displacement, *b)* base soil failure; *c)* tilting (overturn); *d)* ground failure

Design methods for gravity retaining walls are described in detail in the literature. Here, we restrict ourselves to plane strain conditions—in assuming three kinematic modes—whereby safety against displacement, foundation failure and tilting (overturning) are examined. Traditional methods based on the equilibrium equations can be extended for earth-made retaining structures without difficulty. However, definitions of the forces and boundary conditions influencing the equilibrium state need careful consideration. We leave the details of this question to the forthcoming sections, but to emphasize the importance of this remark a convincing simple example is shown below:

The gravity wall shown in Fig. 4–3 retains an earth mass which is reinforced with lateral strips, but the strips are *not* connected to the wall. The reinforced solid mass significantly reduces the earth pressure acting on the retaining wall which should be, therefore, designed for an earth pressure which is less than the "at rest" value derived from parameters relating to the undisturbed soil. Of course, to determine the real value of the decreased earth pressure acting on the wall is by no means a simple problem.

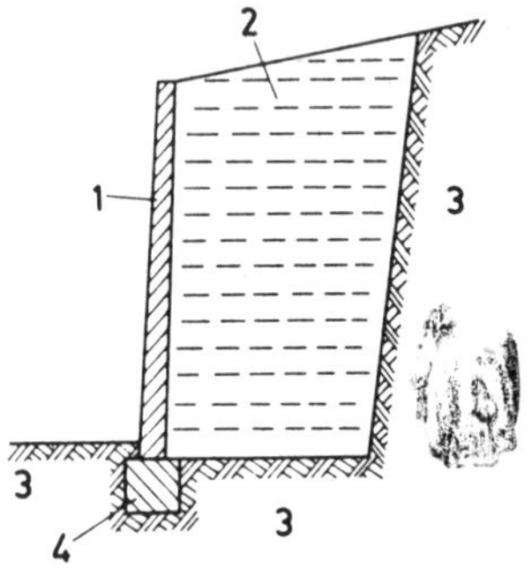

Figure 4–3 Earth pressure reduction by reinforced backfill

1 concrete retaining wall; *2* reinforced backfill; *3* stiff clay; *4* foundation

4.3.2 Investigation of internal stability

All equilibrium and compatibility investigations that guarantee the stability of the soil–reinforcement composite regarded as an anisotropic block, are classified as internal stability investigations. It has to be shown, in general, that

— the reinforcement (or its elements) can bear the stresses induced by the external loads,
— the conditions for proper soil–reinforcement interaction are fulfilled,

— the facing elements have appropriate bearing capacity,
— the probable kinematic failure modes (pull-out, protrusion, etc.) are excluded.

The variety of earth wall constructions strongly influences the form of the internal stability analysis. This question will be reexamined again in details in Chapters 5 and 6.

4.4 GEOTECHNICAL CONSIDERATIONS IN RESPECT OF EARTH-MADE STRUCTURES

Up to now, we have tacitly assumed that the general character and the constructional details of earth-made retaining structures are determined when the geometric requirements are satisfied. Now we have to extend this assumption, since other geotechnical conditions, such as local circumstances (soil properties, stratification, groundwater, etc.) may have a significant role in decision-making. The advantages of these circumstances should be exploited of course.

In addition to considerations connected in general with retaining wall construction (analyses of questions of soil stratification, bearing capacity, drainage, etc.) properties of locally available soils must be examined in more detail. Compactibility, properties connected with stabilization or grouting technologies, suitability for boring, trenching, corrosiveness, etc. should be taken into account at the start when the soil exploration programme is planned. Without these investigations, a soil mechanics expert's report will be insufficient to arrive at an economic (even safe) design. We refer the reader to Fig. 4–3, where no reinforced earth-pressure decreasing operation would be required if the undisturbed soil mass could be grouted or nailed. On the other hand, an optimal arrangement could be achieved if the backfill consists of a stiff clayey mass, or a granular soil can be found for construction purposes near to the site.

5

HORIZONTALLY REINFORCED EARTH RETAINING WALLS

The analogy in *Vidal*'s striking example (*1972*) between the way in which stable sandy and gravelly coastal banks are woven through by pine-needles and reinforced earth retaining walls, was much misunderstood because the former relies on randomly oriented reinforcement whereas the latter consists of laying steel strip reinforcement in a systematic manner. However, this analogy illustrates the basic principle of interaction between the two structural elements: the tensile force produced by the shear stresses between the two interacting materials interface. There was a need for such an illustration especially when the technique was first applied because some experts viewed the technology with scepticism.

The first reinforced earth retaining wall was erected in the Pyrenees (Pragneres) in 1964, but the first paper was published by the inventor two years later (*Vidal, 1966*). Having built some other structures, the characteristic cruciform-shaped precast concrete panels were patented in France and abroad in 1969 (Fig. 5–2*b*). Widespread application of these walls started in the beginning of the seventies and since then dramatic world-wide success has been achieved (*Petrik, 1977*).

By 1970, about a hundred structures had been erected, some of them with a length of many kilometres; the maximum height being 25 m. In the early 1970's 'Reinforced Earth Companies' opened in several big cities (Paris, Washington, Montreal). By 1975, about one thousand, and by 1977, fifteen hundred retaining walls were standing with face area of 1.2 million m^2. By 1982, 5000 had been built demonstrating the general success of the technique (*Gedney, McKittrick, 1975*; *Warwick, 1982*).

Reinforced earth retaining walls have been built in Japan since 1968, in the USA since 1973, in Spain and in Germany since 1976 (*Bongartz, 1976*). In England a long patent debate delayed experimental construction (developments were published in *Civil Engineering*, Vol. 1975–1976). But after this, however, a modified technique, the York system was developed there (*Jones, 1978*). Reinforced earth retaining walls have at this time been constructed in some forty countries.

After the first publication in Hungary (*Kézdi, 1966*) a description of a reinforced structure and its design was presented by the authors (*Scharle, Szalatkay, 1977*). The Hungarian experimental wall was built in 1978, at Fényeslitke (*Fejér, Szalatkay, 1978*). Some other experimental walls were designed in the late 1970's but these were not built thus demonstrating investors' and contractors' aversion to them.

In the mid-1970's a new reinforcing system was developed. Having fabricated geotextiles and geogrids in a large quantity, the idea of using them for earth-retaining walls was proposed. So far, however, only temporary and less significant walls have been reinforced using these materials. However, in the future, worldwide application of sheet-like reinforcements for important and demanding structures may be expected.

In this chapter, the most important characteristics of these structures including their mechanical behaviour, design and construction are presented.

5.1 STRUCTURE AND MECHANICAL BEHAVIOUR OF RETAINING WALLS REINFORCED IN HORIZONTAL LAYERS

These walls act as gravity walls and consist of soil layers and sheet-like reinforcement laid on each other in a systematic order. The facing is usually vertical or very steep and it forms a cover and stabilizes the free end of the wall. It is connected to the reinforcement and can be made of the same material. The soil used for the wall does not differ from that used to infill behind it (Fig. 5–1).

The three component parts of the reinforced earth retaining wall are the reinforcing inclusions, the soil and the skin (or facing). The form and material of the structural elements can be varied within broad limits. Reinforcement can be made of metals (steel, aluminium, or steel), plastic geotextiles or geogrids, in the form of strips or sheets. In selecting the appropriate material, the costs of protection against corrosion and the required durability should be taken into account. The price of stainless or coated steel is higher than that of common steel. Aluminium is also subjected to point corrosion and its tensile strength is less than that of steel. The disadvantage of metal strips is that the frictional resistance on the interface between the strip and the soil is very low and so the linear interaction between the two materials is also small. For other materials the determination of their short term and long term tensile strengths has proved to be problematic.

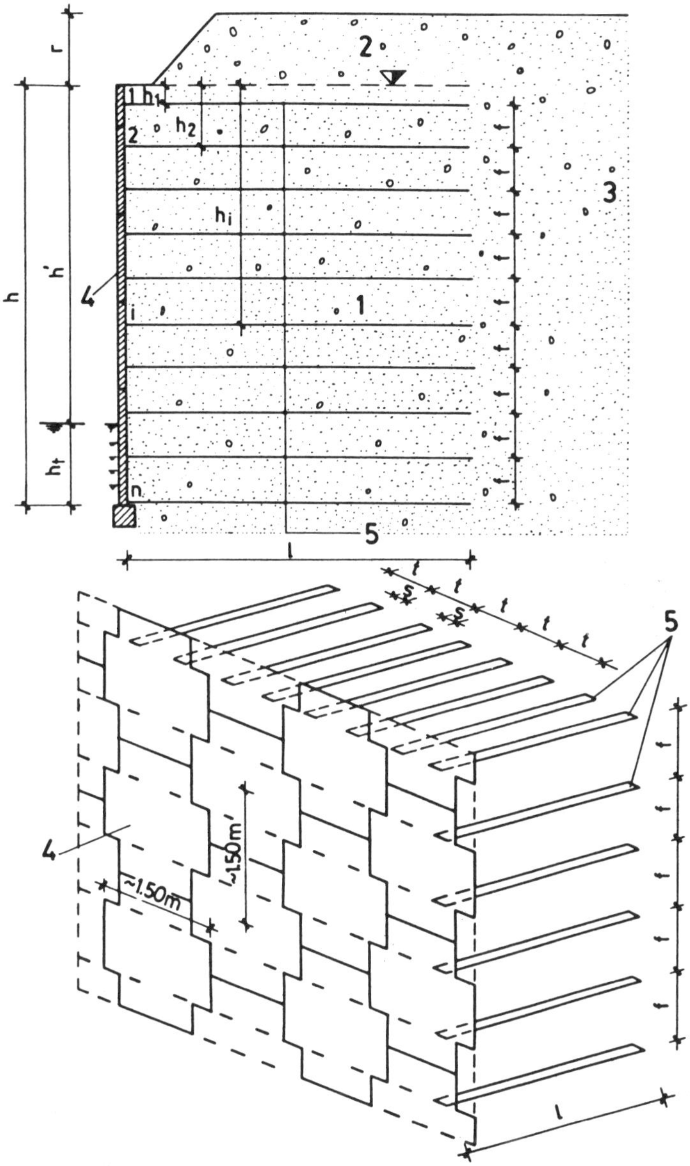

Figure 5–1 Component parts and primary dimensions of the reinforced earth retaining wall

1 retaining wall; *2* fill; *3* backfill; *4* facing; *5* strips

h total height of the wall; *h'* free height; h_t foundation depth; *r* height of the upfill;

f vertical spacing of the ties; *t* horizontal spacing the ties; *s* the width of the ties;

l length of the ties, i.e. width of the retaining wall

Granular materials are usually preferred; the greater the angle of internal friction of the material the more useful it is for the backfill. The facing or skin elements can be made of precast concrete panels, metals or plastics. Many forms (cruciform, hexagonal, H-formed etc.) and surface finishes can be used to give different aesthetic appearances. In addition to the main structural elements, other supplementary parts (connecting bolts and fittings, finishing and corner elements) also have to be used for reinforced earth retaining walls. Naturally, foundations and draining systems also form part of the whole structure.

Until now two types of reinforcing systems have been usual. Strips reinforce the backfill material along distinct lines whereas sheets reinforce it through the whole cross-section of the wall along one continuous surface (Table 5–1).

Table 5–1 Retaining walls with reinforcing layers

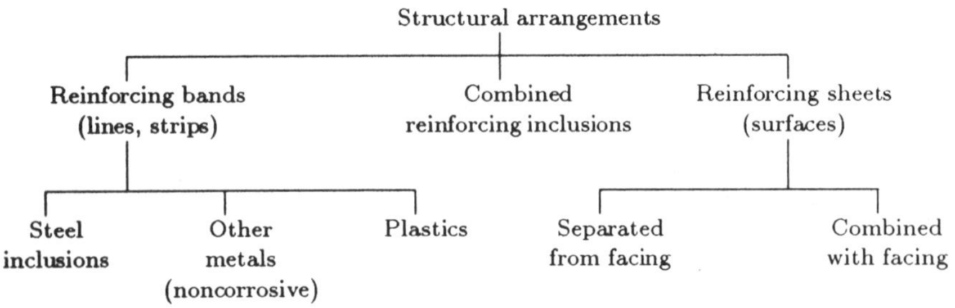

However, there are also examples of rod, and bar reinforcement: they behave rather like anchors and are, therefore, omitted from our discussion. As for the facing elements, rolled steel plate facing, preferred initially, has been superseded (Fig. 5–2*a*), mostly because of the high cost of the material and its tendency to corrode. The Vidal-type cruciform concrete panel, its many variations (Fig. 5–2*b*), and the hexagonal concrete panel of the York system have been widely applied (Fig. 5–2*c*).

For earth retaining walls reinforced by sheet (or fabric) reinforcements, it is advantageous, after laying them onto the surface of the soil, that they can be wrapped around the layers of the rolled earth backfill (Fig. 5–3). If the fabric on the surface prevents the soil from running out from between the fabric layers, then rigid concrete or steel members are not necessary for the face of the structure and reinforcement can serve both as an earth anchor and as a wall facing. In cases like these, even the connecting elements become unnecessary. Moreover, sheet-like reinforcements decrease the earth pressure even if they are not attached to the facing (*Holtz, Broms, 1977*); some of the steel strips in a conventional reinforced earth retaining wall can be substituted for by them. Besides

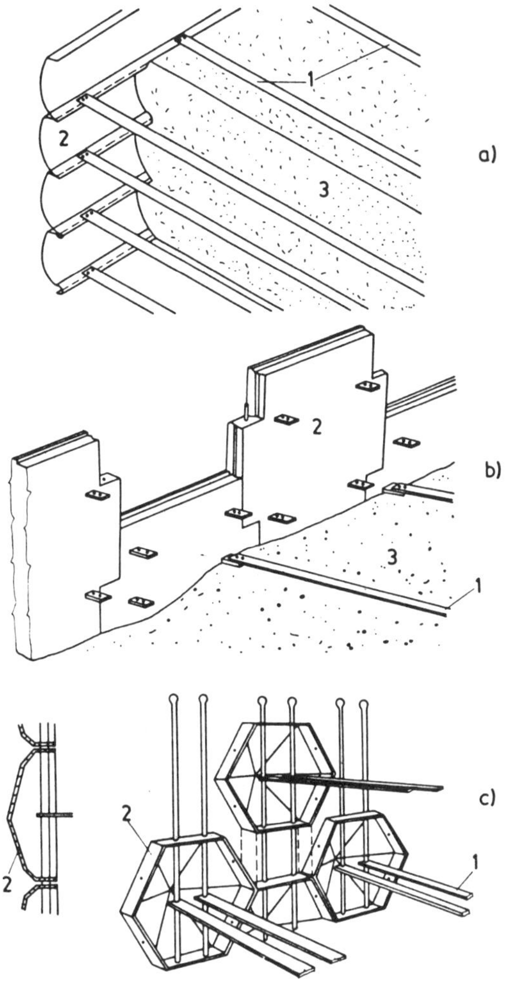

Figure 5–2 Construction method for the reinforced earth retaining wall
a) hemispherical steel skin elements; *b)* French-type cruciform shaped precast concrete elements; *c)* construction of the York system reinforced earth retaining wall
1 strips; *2* facing; *3* backfill

decreasing the quantity of steel, this arrangement can be advantageous particularly in the upper zone of the wall (*Szalatkay, 1981*) where the corrosion hazard is high and vertical pressure is small.

Here, two mechanical effects of reinforced earth retaining walls detailed in Chapter 3 should be emphasized. The first one is the need for frictional resistance on the interface between the soil and the reinforcement and thus the magnitude of the normal stress is important.

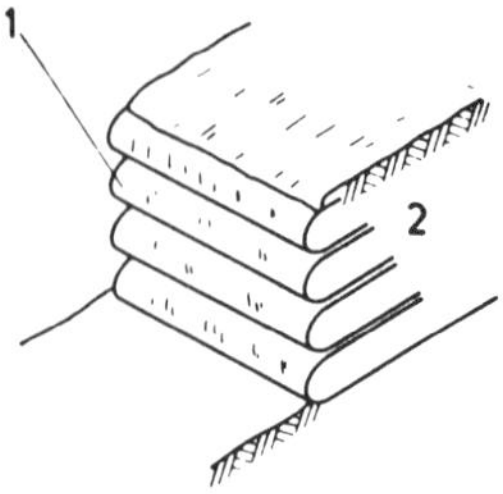

Figure 5–3 Fabric sheet reinforced earth wall

1 sheet-like reinforcement; *2* soil

The second most important requirement is adequate tensile strength of the reinforcement.

The normal stress can be calculated from the mass of the soil, but it may be increased by various additional loads acting on the surface of the backfill. The tensile force in the reinforcement is:

$$T_{\mathrm{a}} = \int_0^A \mathrm{d}T = \int_0^A \sigma_1 \tan\delta \, \mathrm{d}A \qquad (5\text{–}1)$$

where $\mathrm{d}A$ = elemental surface of the reinforcement contacting the soil, σ_1 = normal stress, δ = angle of friction between the soil and reinforcement.

The interaction between soil and reinforcement is the result of the resistance acting on the contact surface. Two kinds of deformation may occur:

— in the soil surrounding the reinforcement, strain arises due to the effect of deadweight and external load. This strain is, however, balanced by the reinforcement by developing tensile strains in it (along the active length of the reinforcement),

— tensile forces cause strain in the reinforcement; in the surrounding soil only a small strain would be developed by other causes; in these cases, the force is transferred from the reinforcement to the soil (passive length).

Theoretically, it is possible, but very rare that the interaction between the reinforcement and the surrounding soil becomes weak within an extended area, because their deformations are equal. (In such cases the reinforcements are unsuitably situated.) Attention should be paid to the fact that the angle of friction and its numerical value in the case of nets, grids, geotextiles and fabrics is the result of complicated interactions and it obscures the details. For example, embedment of the reinforcement, unevenness of soil surface all influence the value of the frictional resistance; we shall come back to this question later. At any rate, it should be noted that the upper limit of the tensile force can be estimated from analogous variations of Equ. 5–1. Naturally, the tensile force cannot exceed the limiting value determined by the tensile strength of the reinforcing material.

From the structural and mechanical points of view, sloping reinforced earth retaining walls can hardly be distinguished from reinforced slopes or embankments. Categorization becomes obscure particularly in the case when the latter are covered, or the facing is left out from the former. We encourage the reader not to dwell over classifying them into categories, since there is no essential difference between them.

5.2 THE REINFORCED EARTH RETAINING WALL

Earth retaining walls reinforced by metal strips are called reinforced earth retaining walls. As mentioned earlier, the terms: reinforced soil, retaining wall and reinforced earth retaining wall are equivalent from a technical point of view. Here, however, only the last expression is used.

Before describing the details of some structures, design and methods of construction, some basic concepts are summarized and some interesting construction examples are presented.

5.2.1 Constructional, mechanical and economical performance of reinforced earth retaining walls

The material used for strip reinforcement in reinforced earth retaining walls is generally steel and its width ($s = 40 \ldots 200$ mm) is much greater then its thickness ($v = 1 \ldots 6$ mm). Strips are positioned relatively densely both in the horizontal and vertical directions ($t, f, = 0.3 \ldots 0.8$ m, Fig. 5–1). A sloped fill can be placed onto the surface of the wall. The dimensions of retaining walls, strips and spacing (Fig. 5–1) are described by the following parameters:

— $h =$ total height of the wall,
— $l(= 0.8 \ldots 1.2\,h) =$ width of the wall or the length of the strips (ties),
— $i = 1 \ldots n =$ the number of reinforcing layers (counted from the top of the wall),

— h_t = foundation depth (for the horizontal surface usually $h_t = 0.1\,h$; for sloping surface $h_t = 0.2\,h$),
— r = height of the fill,
— h' = free height of the wall,
— $h_1, h_2, \ldots h_i$ = vertical distance of the strip levels measured from the top of the wall,
— $t, (t_1, t_2, \ldots t_i)$ = horizontal spacing of the strips for each level,
— $f, (f_1, f_2, \ldots f_i)$ = vertical spacing of the strips,
— $s, (s_1, s_2, \ldots s_i)$ = width of the strips for each level,
— $v, (v_1, v_2, \ldots v_i)$ = thickness of the strips for each level.

The height of reinforced earth retaining walls is theoretically unlimited. Economical advantages become evident above heights of $h = 3.5 \ldots 4.0$ m. The highest wall built worldwide is 40 m in Pakistan (*Smith, Birgisson, 1979*).

Not only retaining walls but also embankments for railways and highways, and bridge abutments have been made from reinforced earth. This construction method can be applied both for sea and river walls, although the loads and stresses must be carefully considered in these cases. Depending on whether the foundation is combined with bridge abutments constructed of reinforced earth, or the bridge base is placed on the surface of the reinforced earth construction, on a pile or column built for this purpose, three fundamental cases can be distinguished (Fig. 5–4).

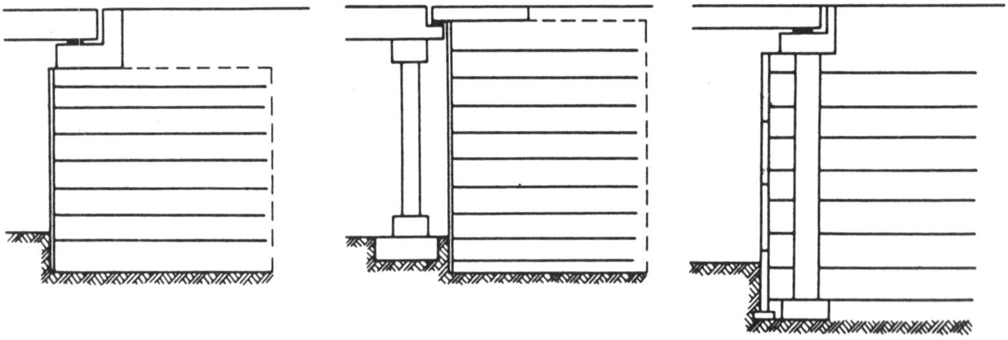

a) b) c)

Figure 5–4 Reinforced earth abutments

a) foundation placed on the reinforced earth structure;

b) foundation separated statically and functionally from the reinforced earth structure;

c) foundation separated statically and built into the reinforced earth structure

In the case where the base is placed on the surface of the reinforced earth bridge abutment, a statically determined construction must be applied and the load on the foundation must be regarded as a partially distributed load. In the case of statically indeterminate construction, settlements developed under the foundation must also be considered during design.

In a hilly countryside or where higher earth embankments are to be supported, twin structures and stepped walls may be erected (Fig. 5–5). Thereby, the aesthetic appearance and mechanical behaviour of the retaining wall can be improved.

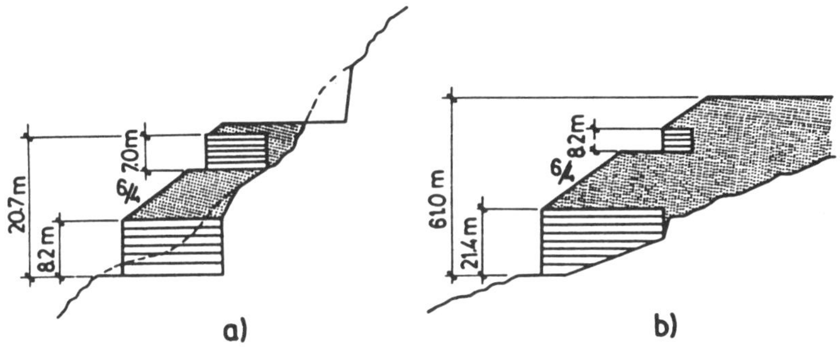

Figure 5–5 Stepped or terraced retaining walls

a) Vigna (France); *b)* Le Peyronnet (France)

The length of the ties for a reinforced earth retaining wall in general varies between

$$l = 0.8 \ldots 1.2h \qquad (5\text{–}2)$$

Both sides of high and narrow road embankments can be supported by reinforced earth retaining walls. Three cases can be distinguished from each other, depending on the location of the free ends of the strips in such double-faced reinforced earth embankments. The strips may be placed

— far from each other,
— just touching at their free ends,
— overlapping each other (see Fig. 5–6).

It must be emphasized, however, that in the last two cases there is no question of the structural connection of the strip-ends, or anchorage.

Reinforced earth retaining walls may be classified economically, technologically, aesthetically or mechanically. According to a complex economical comparison (*Lee, Adams, Vagnernon, 1973*), the approximate breakdown in cost varies depending on the height of the wall (Fig. 5–7). Economic efficiency of an earth retaining wall reinforced by steel strips is influenced by the prescribed durability and corrosion rate of the steel, which depends on the surrounding soil. It can be stated in general, that

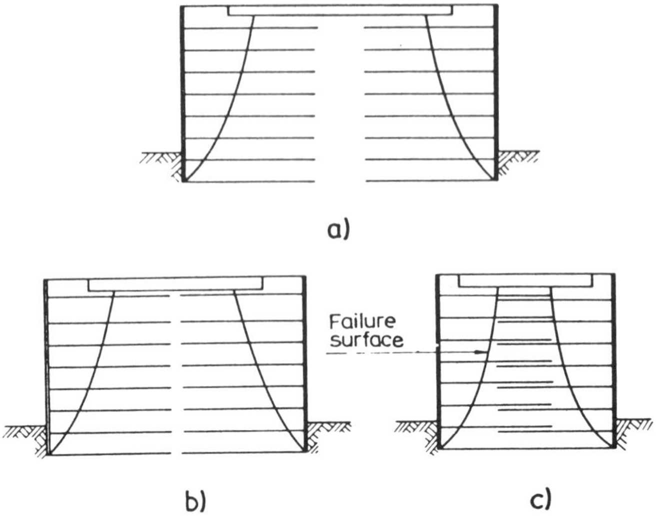

Figure 5–6 Double facing reinforced retaining walls

a) ends of the strips are far from each other;

b) ends of the strips connected to both sides of the facing elements reach each other;

c) strips overlap each other

even in the case of increased corrosion hazard, a reinforced earth retaining wall is cheaper than any of the conventional walls. In ideal conditions they might be expected to cost about half as much as cantilever or crib walls of the same height.

For higher walls, the relative saving increases. The economic efficiency is mainly achieved through the quantity of building materials used for the construction: the need for cement is expected to be about one tenth of the quantity used in gravity walls and a half or a third of the material used in cantilever walls. The quantity of steel, too, decreases compared to that used for reinforced concrete walls.

In Fig. 5–8 and Table 5–2 the relationship between the height of a wall and the relative cost is demonstrated (*Păunescu, Donca, 1979*).

An economic analysis has shown that a 6 m high reinforced earth retaining wall (reinforced concrete facing panel, stainless steel strips) provides at least a 20–25% saving as compared with gravity or cantilever walls of the same height. This result is in accordance with other cost estimations (*Szalatkay, 1975*) and it confirms the economic efficiency of these structures regardless of the different bases for the cost calculations.

In this comparison, there is special weight for economic factors such as simpler construction, the possible use of local soil materials, and the absence of form-work. The great economic advantage of reinforced earth retaining walls lies in their mechanical behaviour because they do not

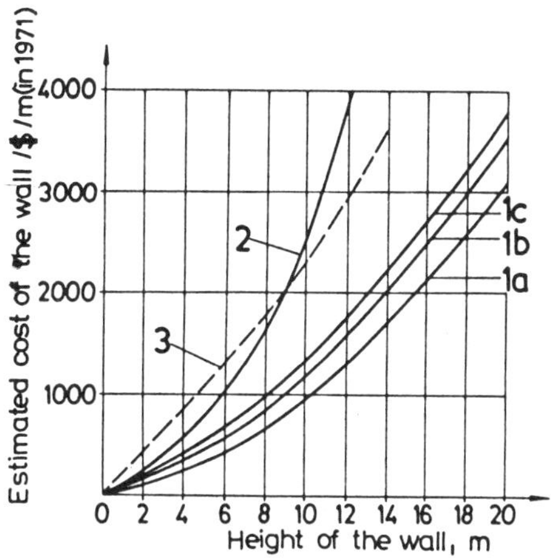

Figure 5–7 Comparative cost analysis for various types of retaining walls (50 years service life). Safety factors for stability analysis 1.5; for precast concrete skins: low walls 10% more, high walls 35% cheaper (*Lee, Adams, Vagnernon, 1973*)
1a) reinforced earth retaining wall, no corrosion;
1b) reinforced earth retaining wall, corrosion rate 0.025 mm/year;
1c) reinforced earth retaining wall, corrosion rate 0.50 mm/year;
2 cantilever wall; *3* metal bin (galvanized metal)

Table 5–2 Dimensions of gravity wall and cantilever wall (m) (interpretation of symbols in Fig. 5–8.)

to *a)*					to *b)*				
H	3.00	6.00	9.00	12.0	H	3.00	6.00	9.00	12.00
h	0.80	0.80	0.90	1.20	$a = 0.25\,H$	0.75	1.50	2.25	3.00
a	0.70	1.05	1.50	1.95	$b = 0.1\,H$	0.30	0.60	0.90	1.20
b	1.05	1.80	2.60	3.50	$c = a$	0.75	1.50	2.25	3.00
c	1.05	3.10	5.00	6.70	$B = 0.6H$	1.80	3.60	5.40	1.20
d	1.15	2.10	3.10	4.10	$h = 0.1\,H$	0.30	0.60	0.90	1.20
e	1.00	1.15	1.40	1.90	$f = 0.0375\,H$	0.11	0.22	0.34	0.45
					$v = 0.05\,H$	0.15	0.30	0.45	0.60
					h_7	0.80	0.80	0.90	1.20

require special foundations, piles, etc., and are not sensitive to unequal settlements. They are economical in that they can be built on soft subsoils without deep foundations (*Walkinshaw, 1975*; Fig. 5–9). In the calculated 25...60% saving in cost, not only savings in building materials but also cheaper construction costs are included. Labour costs on building site is greatly reduced, since the major requirement is the assembly of prefabricated elements.

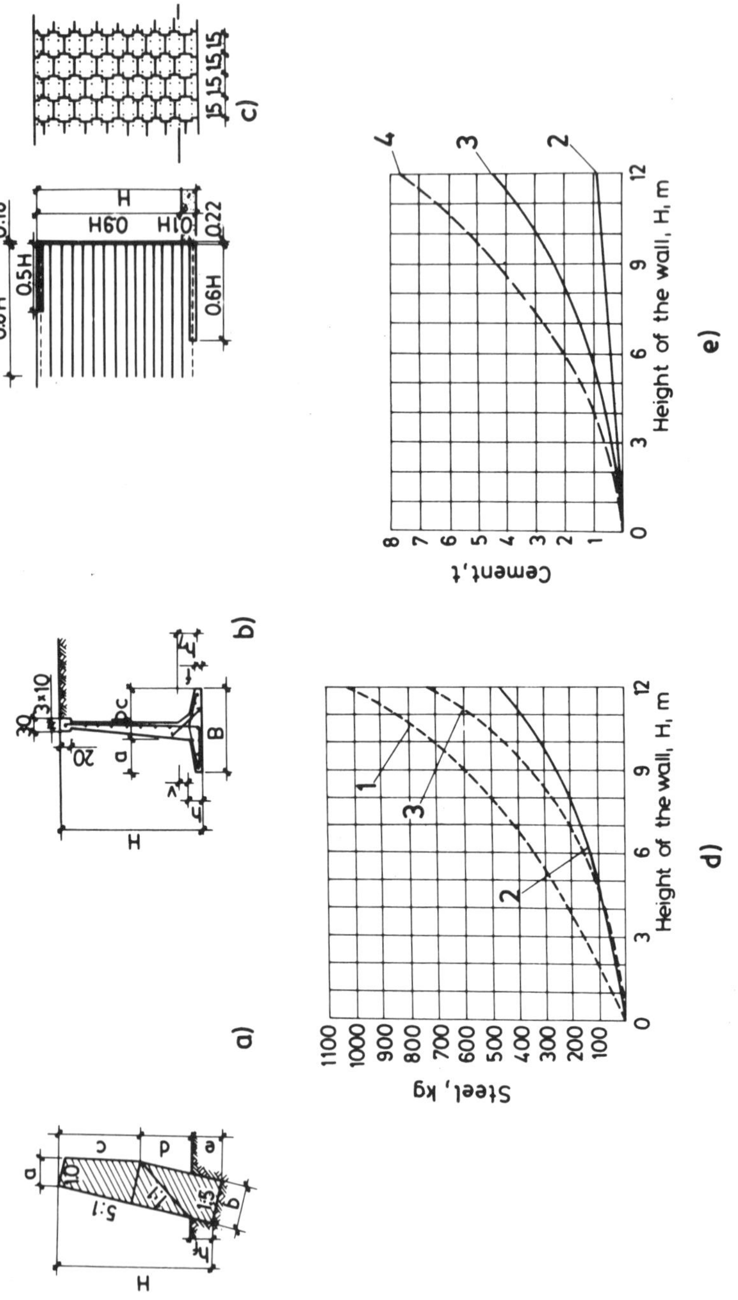

Figure 5–8 Material consumption for gravity, cantilever and reinforced earth retaining walls (*Păunescu, Doncă, 1979*)

a) gravity wall; *b)* cantilever wall; *c)* reinforced earth retaining wall; *1* metal skin; *2* with precast concrete facing element; *3* cantilever wall; *4* gravity wall

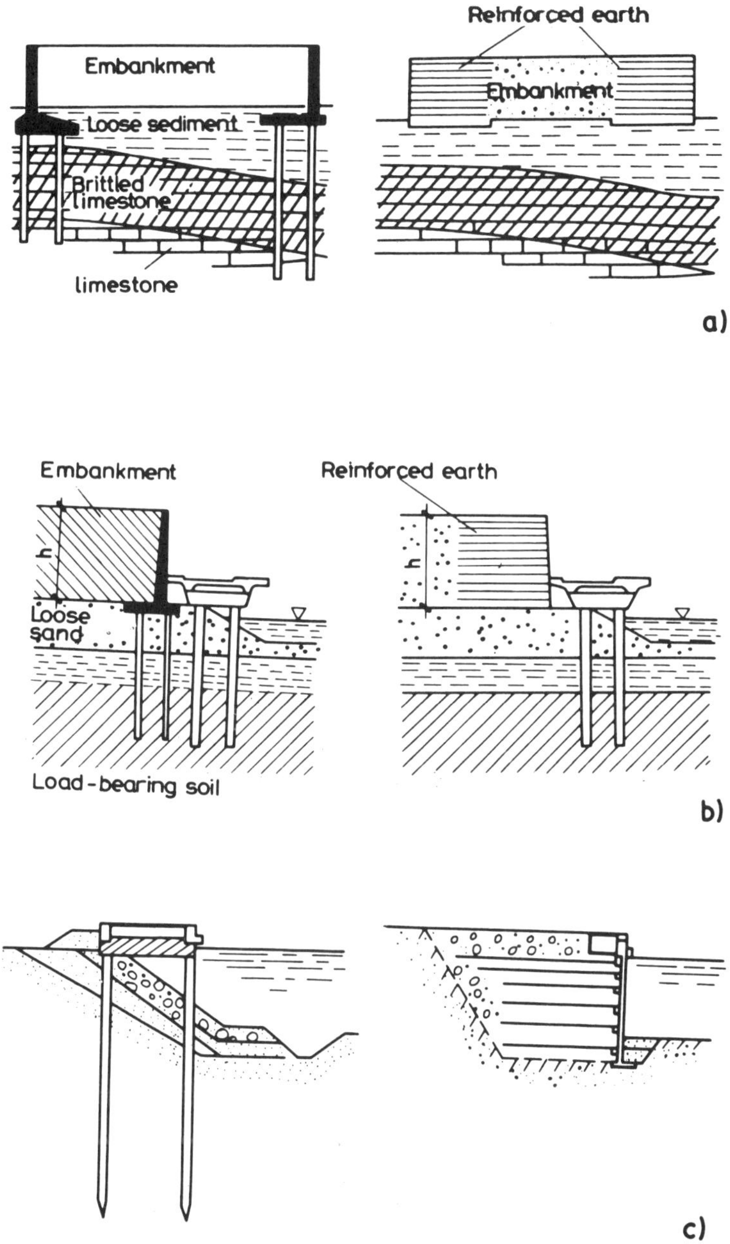

Figure 5–9

a), b), c)

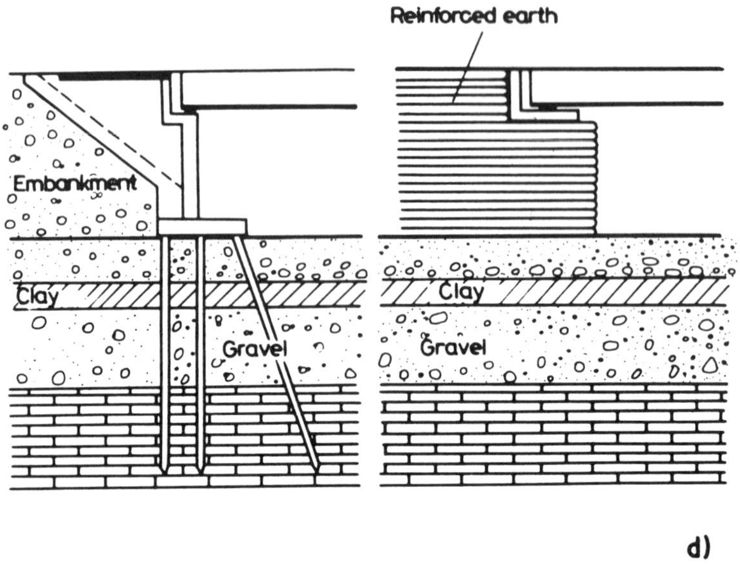

Figure 5–9 Elimination of deep foundations (*Walkinshaw, 1975*)
a) Bridge de Sèvres (1971)—saving 45%; *b)* Sète (1971)—saving 50%;
c) Valras (1971)—saving 15–20%; *d)* Thionville (1972)—saving 60%

Any general contractor can construct a reinforced earth wall since no specialized experience is required. The concrete panels used for facing have an approximate weight of 10 kN; only a light crane and 2...4 unskilled persons are necessary to transport the panels and lift them into place. Assembling the strips and the facing panels is very simple and can be carried out at the same time as the backfill is placed. In the assembling cycles, the earth work—placing and compacting—can easily be mechanized. Facing elements are prefabricated so that any forms and facing finishes can be developed (Fig. 5–10).

By using articulated facing elements and erecting them with a terraced and combined slope–earth wall cross-section, a pleasing environmental form can be obtained (Fig. 5–11).

For the sake of completeness the disadvantages of the construction technique should also be mentioned i.e. a great quantity of well compactable granular material is required and because of that, for lower ($h < 4$ m) walls, where relatively more steel material is needed, the structure is expensive. In this range of dimensions, an economic analysis is necessary to decide whether or not it is worthwhile transporting fill material to the site in order to build a reinforced earth retaining wall.

Figure 5–10 Saint Joseph Chapel in Montreal (Canada)

Figure 5–11 Stepped, curved reinforced retaining wall (Paillon, France)

5.2.2 Examples of reinforced earth retaining walls

Before describing the structure, mechanical behaviour and economy of reinforced earth retaining walls, their general character have to be discussed because the examples built to date differ widely. The efficiency of typical construction systems is illustrated by a few examples from the wide choice available.

Figure 5–12 Curved reinforced retaining wall (Cove Point, USA)

A retaining wall, of free height $h' = 4.5$ m, built in Cove Point (USA, Columbia) is a good example of a curved layout (Fig. 5–12). The rectangular cylindrical facing units of the reinforced earth retaining wall in Vail Pass (USA) have an attractive and dynamic appearance (Fig. 5–13). This latter structure is a good example of adapting the wall to the prevailing environment by erecting terraced or stepped walls on the hill-side.

Figure 5–13 Stepped wall with an aesthetic facing (Vail Pass, Colorado, USA)

The reinforced earth retaining wall erected in Revigny–sur–Ornain has complex functions; it works not only as a bridge abutment but also as a pillar in France (Fig. 5–14). The bridge as well as the wall are curved.

A reinforced earth coast wall which supports a road embankment is shown in Fig. 5–15. It is more than 300 m long, 8.5 m wide and 7.5 m

Figure 5–14 Abutment and pillar from reinforced earth (Revigny–sur–Ornain)

high. Loads acting on the wall are considerable: 2 m changes in the water level (tidal changes) and a railway on the surface of the fill exerts an additional load (*Walkinshaw, 1975*).

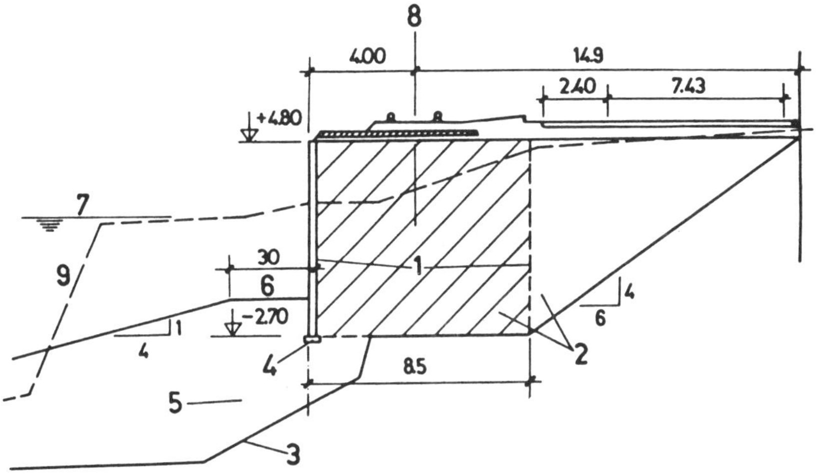

Figure 5–15 Coastal reinforced earth retaining wall (Georgia, Newcastle street, Brunswick; *Walkinshaw, 1975*)

1 reinforced earth retaining wall; *2* backfill; *3* boundary of the instable cut; *4* concrete foundation; *5* embankment; *6* bench; *7* yearly high water level; *8* railway; *9* original ground

According to *Jones* and *Bolton* (*1977*): "The development of reinforcement systems should be directed towards the cheapest materials possessing the greatest strength, good surface friction, and resistance to corrosion. Although widely used to date, galvanized steel is susceptible to corrosion. Stainless steel is expensive. Aluminium might provide corrosion-free service but has a relatively high cost/strength ratio." A promising new material, in Dr. Jones' opinion, is glass fibre reinforced plastic (GRP).

In the last decade, new techniques and materials have contributed to further development of such structures. *Al-Hussaini* (*1977*) described an earth retaining wall reinforced by nylon strips coated with neoprene. The height of the wall was 5 m approximately, the horizontal and vertical spacings were $t = 1.20$ and $f = 0.60$ m, respectively. The length of the ties was 3 m, their thicknesses 3 mm, and their widths 100 mm. The backfill was sand ($\phi = 35°$). The construction differed from those built in general in the material of the strips but the way in which they were connected to the facing elements was similar (Fig. 5–16).

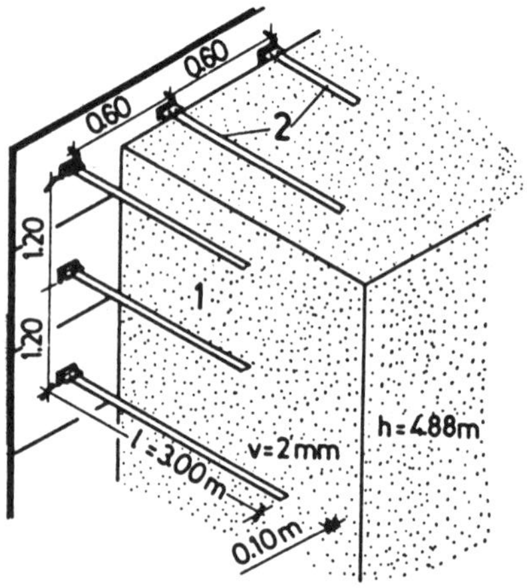

Figure 5–16 Reinforced retaining wall according to *Al-Hussaini*

1 soil; *2* nylon reinforcing strips coated with neoprene

In Hungary, between 1978 and 1981, three retaining walls—two of them experimental—were built, and two others were designed, but never constructed. The one at Fényeslitke (a village in North-Eastern Hungary, Fig. 5–17) is 4.5 m high and 18 m long, and supports a railway embankment. The backfill is made of local soil material, a silty sand, with uniformity coefficient $U = 10.8$; mean particle size $d_m = 0.035$ mm, maximum dry density $\rho_{0,\max} = 1.89$ g/cm^3.

The cruciform facing elements were prefabricated; the reinforced concrete elements and their dimensions were designed by the method of *Schlosser* and *Vidal.* The horizontal and vertical spacings between the ties are 0.75 m and the strips are 3.0 m long, 4 mm thick and 0.10 m wide. The wall was finished within three days (*Fejér, Szalatkay, 1980*).

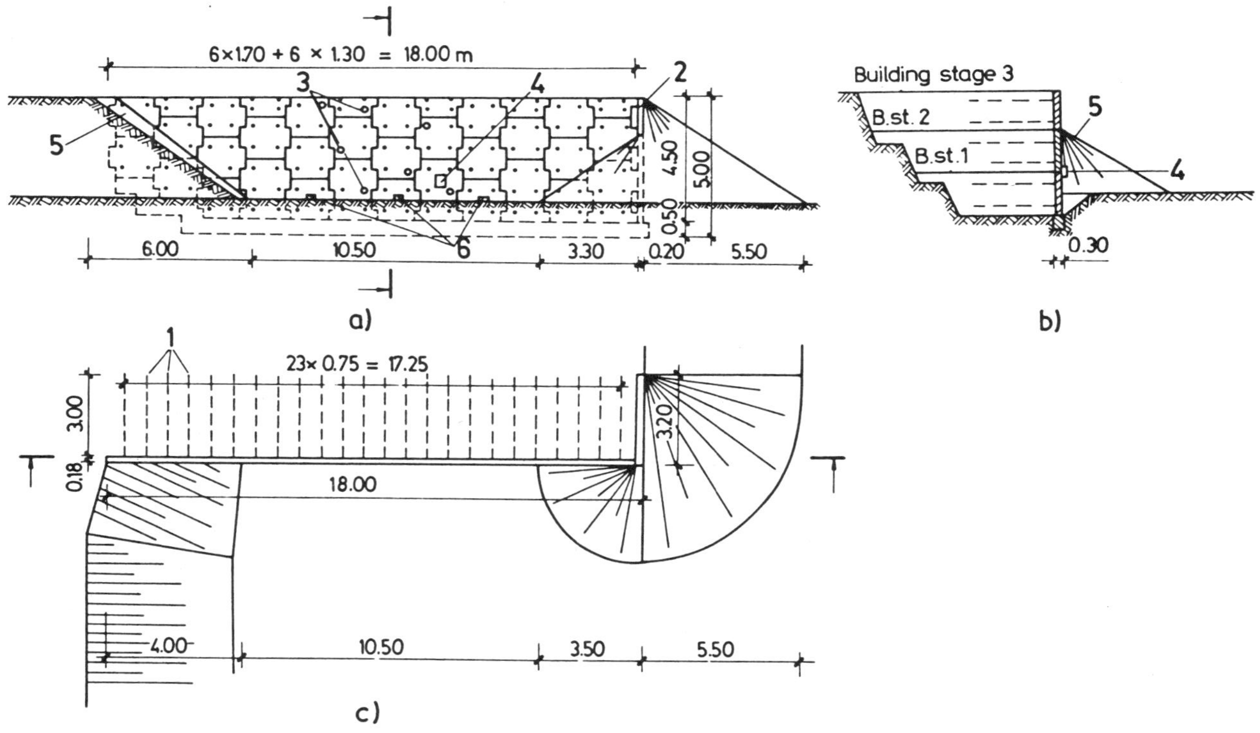

Figure 5–17 Front and cross-sectional view of the reinforced earth retaining wall at Fényeslitke (Northern Hungary)

a) section A; *b)* section B; *c)* layout plan

1 steel strips; *2* closing reinforced concrete element; *3* measuring strips; *4* measuring box; *5* slope; *6* drains

The height of the two walls in Siófok (South-Western Hungary) varies between 3.2...5.2 m, both supporting a ramp of the roadway to the bridge spanning over the river Sió (Fig. 5–18). The total length of the walls is 120 m, the trace of the facing is followed by a horizontal curve of radius $R = 1500$ m. The length of the strips varies between 3.1...3.6 m, their widths are 120 mm, and their thicknesses are 4 mm. For facing elements the decorated French forms applied in Fényeslitke were used. The elements were prefabricated and the walls were constructed in the summer and autumn of 1980 (*Kerényi, Szalatkay, 1981*).

Figure 5–18 Reinforced retaining wall built in Siófok (Hungary)

The economic advantages of the reinforced earth retaining wall in Siófok can be demonstrated by material consumption indices. A cantilever wall having the same height and length would have required 570 m^3 reinforced concrete cast in-situ. For the reinforced earth retaining wall equivalent to this, only 140 m^3 precast concrete was required. The steel material used in Siófok was a third less than would have been required for the cantilever wall (37 tons), the price of which would have been further increased by the additional cost of about 1600 m^2 form-work and scaffolding, etc.

The strips of the third Hungarian reinforced earth retaining wall were made of polyester reinforced fibreglass to avoid corrosion (*Marczal, 1982*). The surface of the strips was roughened by coarse sand in order to increase their passive resistance against pull-out forces. Using 6.0 m long, and 4×50 mm cross-sectional strips spaced in 0.5 and 1.0 m vertically and horizontally it was possible to erect a 5.4 m ($h = 6.0$ m) high experimental wall (Fig. 5–19).

An individual structure was designed to adorn the environment of a monument (*Szalatkay, 1978*). The wall, a type which had not been built

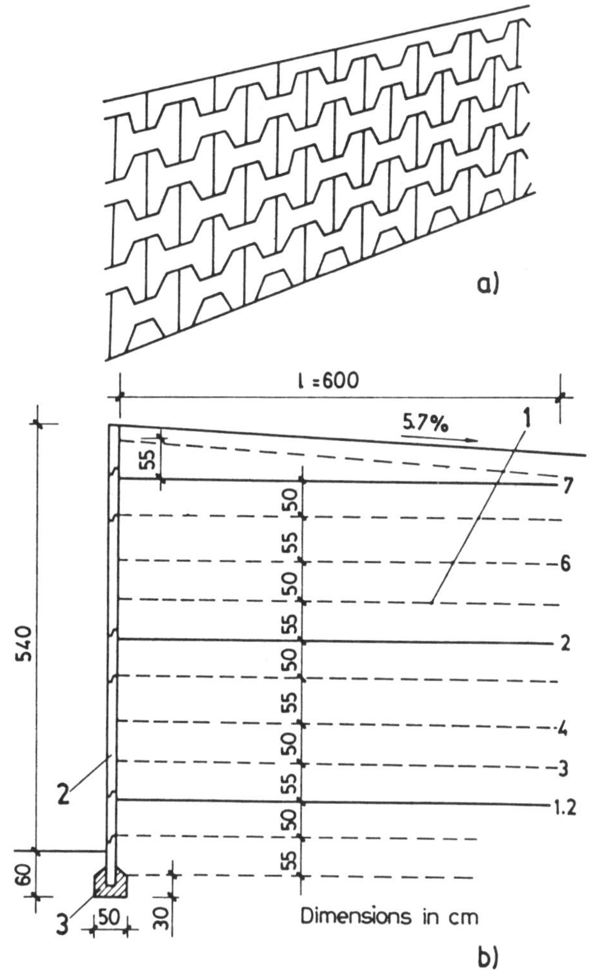

Figure 5–19 Fibreglass reinforced polyester strips and H-shaped facing elements, prefabricated by the Company for Building Concrete Road (*Marczal, 1982*)

a) front view of the facing; *b)* cross section;

1 strips; *2* facing; *3* prefabricated reinforcing concrete found

previously, would have supported an embankment up to 5.5 m high (Fig. 5–20).

The upper 1.0... 1.5 m high monolithic cast in-situ reinforced concrete zone of the facing would have been stabilized by anchoring strips. Below this, two types of facing elements with different functions and ap pearance would have been built. The bigger ones would have been an-

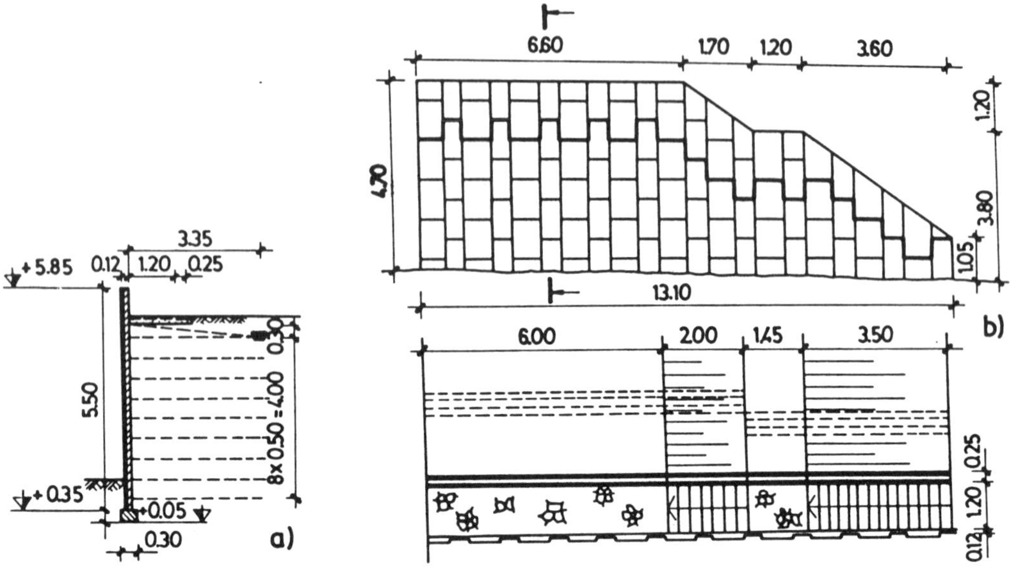

Figure 5–20 Reinforced earth retaining wall designed at Sopron (Western Hungary, *a)* cross section; *b)* front view of the facing; *c)* plan

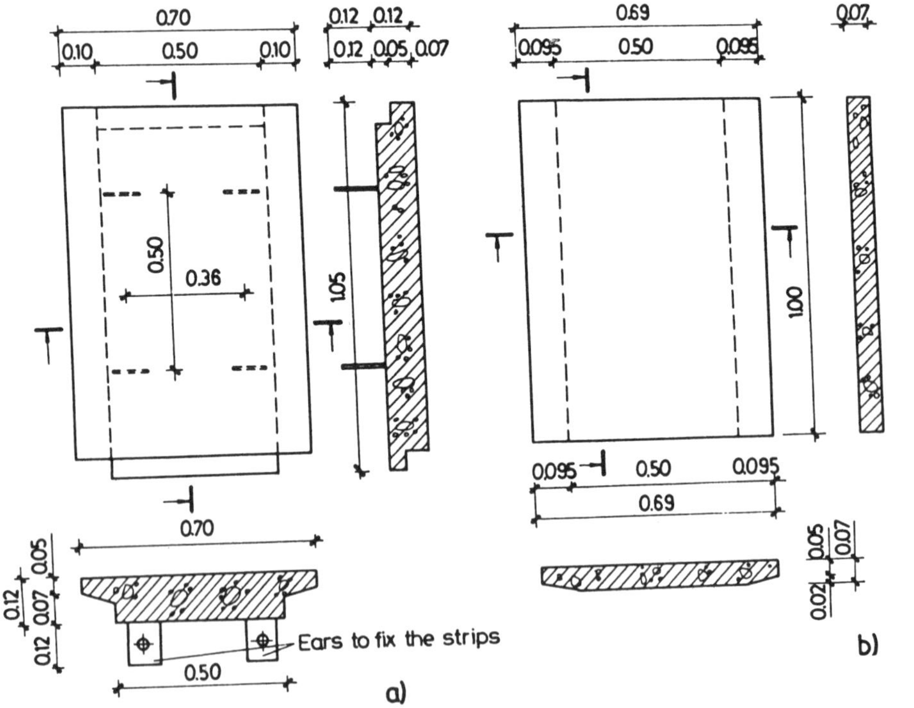

Figure 5–21 Facing elements of the reinforced earth retaining wall designed at Sopron *a)* elements anchored backwards to the fill; *b)* elements fitted between the anchored elements

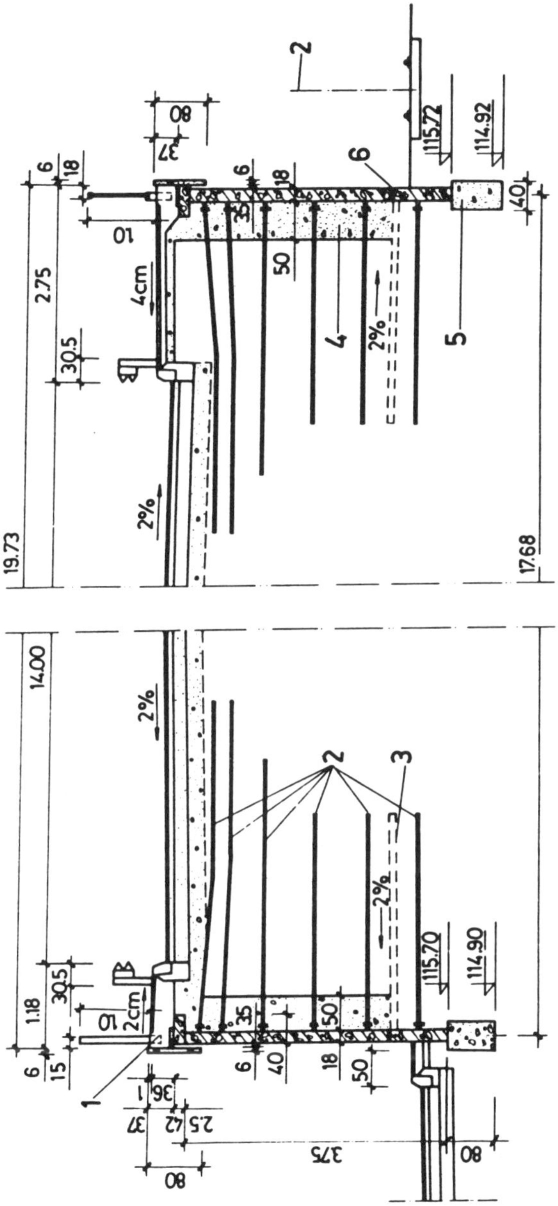

Figure 5–22 Cross-sectional view of the reinforced earth retaining wall designed at Kecskemét (UVATERV 1981)

1 precast reinforced concrete capping; *2* ribbed and galvanized steel strips, 160–3 and 100–3; *3* soil stabilization for drain; *4* drainage layer; *5* strip foundation chored back by four strips; the smaller ones would have been held in place by the neighbouring larger strips and also by the backfill (Fig. 5–21).

On the higher section of the wall, the strips were 3.35 m long and 85 mm wide, on the lower section the strips were 2.80 m long and 100 mm wide and connected to the facing units.

All the advantages mentioned earlier for this structure would have been attained in Kecskemét (Southern Hungary), where a 300 m long and 4.0 m high reinforced earth retaining wall was designed (Fig. 5–22). Economic efficiency would have been expected by prefabricating the great number of elements (about 400) for this representative wall. Here because of the bad soil properties, ribbed steel strips would have been laid into the backfill to increase frictional resistance.

5.2.3 Construction details of reinforced earth retaining walls

In the previous chapter, some general information on the reinforced earth retaining walls was summarized but no specifications were given to identify appropriate materials. The requirements of design (dimensioning and choosing construction type), however, need more extensive analyses. Therefore, in the following paragraphs, we deal with the mechanical, chemical, etc. interactions between soil, strips and facing elements.

5.2.3.1 Frictional resistance between soils and strips

The frictional resistance between soils and strips is a complex physical interaction between the strips and the individual particles and the soil mass. Displacement of an individual particle along a surface can be caused not only by sliding but also by rolling. In addition, passive resistance is affected by surface micro-dimensional unevenness and even by the embedment. During shear deformation and sliding, individual particles move in a direction perpendicular to the surface and the soil mass is rearranged. Resistance developing in this way is called frictional resistance because it is directly proportional to the normal pressure acting on the particles, to a close approximation. The friction can be defined with the help of *Coulomb*'s law and the coefficient of proportionality (μ) is called the frictional coefficient (tangent of the angle of friction).

The frictional resistance between various soils and construction materials can be measured in a shear box (*Potyondy, 1961*), or by the pull-out test (*Lee, Adams, Vagnernon, 1973*).

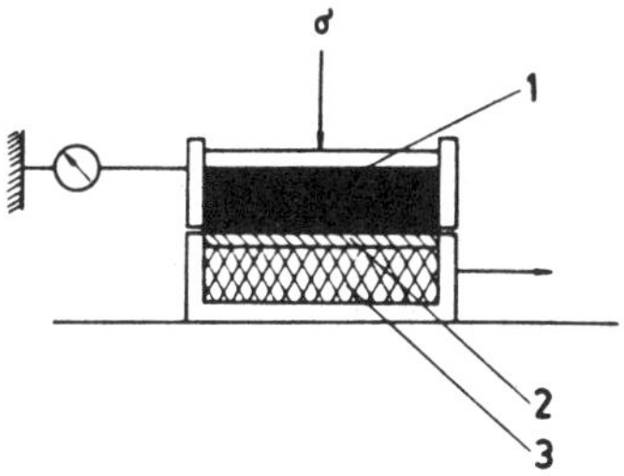

Figure 5–23 Measurement of the frictional resistance between soil and strips in shear box
1 soil; *2* steel strips; *3* supporting mass

The French recommendation for designing reinforced earth retaining walls entails the use of the shear box: the strip, with the same surface properties as that to be built into the wall is placed in the lower half of the box and the soil material is placed in the upper half. The shear strength is measured in the same way as for the soil alone and shearing is carried out at constant speed (*LOTAR, 1979*; Fig. 5–23).

For the pull-out test, a simple apparatus can be constructed (Fig. 5–24; *Fejér, Szalatkay, 1980*). The arrangement corresponds better to the model of a strip embedded into soil, but errors arise from disturbance, mostly resulting from arching effects in the box and this affects the accuracy of the results.

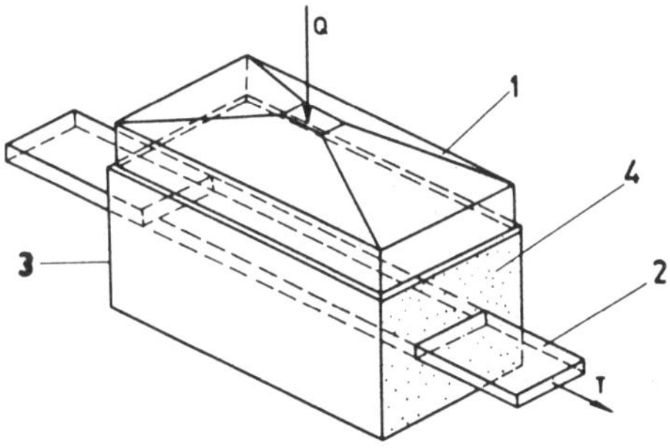

Figure 5–24 Pull-out friction test
1 loading plate; *2* steel strip; *3* soil box; *4* soil

Another test method involves placing inclined reinforcement into the shear box. This method is particularly useful for clays: not only the frictional resistance but also the shear resistance can be obtained depending on the inclination of reinforcement (*Ingold, 1981*).

For the frictional resistance, detailed experimental information is given by *Potyondy* (*1961*). It is known in general that the angle of friction

(δ) and adhesion (c_a) between the soil and steel are influenced by the type and state of the soil (density and moisture content). As the angle δ is proportional to the angle of internal friction (φ), and the adhesion can be defined by the cohesion, the coefficients of proportionality are

$$f_\varphi = \delta / \varphi \tag{5-3a}$$

and

$$f_c = c_a / c. \tag{5-3b}$$

Beside particle sizes and cohesive effects, the roughness of the steel surface is an important factor. On smooth surfaces, the shear resistance is 10...40% less than that for rough (rusted) surfaces. The adhesion coefficient depends on the moisture content; the lower the moisture content the stronger is the cohesion of the soil. The results listed in Table 5–3 can be used for design purposes (*Potyondy, 1961*).

Table 5–3 Frictional coefficients between steel and soil types (*Potyondy, 1961*)

Steel reinforcement	Soil type and state								
	Sand		Cohesionless silt			Mixed soil		Clay	
	$0.06 < d <$ < 2.00 mm		$0.002 < d < 0.06$			50% sand + + 50% clay		$d \leq 0.06$ mm	
	dry	saturated	dry	saturated		$I_c = 1.0 \ldots 0.5$		$I_c = 1.0 \ldots 0.73$	
	dense		dense	loose	dense				
	f_φ	f_φ	f_φ	f_φ	f_φ	f_φ	f_c	f_φ	f_c
Smooth (polished)	0.54	0.64	0.79	0.40	0.68	0.40	—	0.50	0.25
Crude (rusted)	0.76	0.80	0.95	0.48	0.75	0.65	0.35	0.50	0.50

According to recent experiments, the angle of friction is influenced by the magnitude of the normal stress acting on the soil. In-situ investigations at the depth $z < z_0 = 6$ m in a wall showed that the friction resistance between soil and strips is

$$\mu = \mu^* \left(1 - \frac{z}{z_0}\right) + \frac{z}{z_0} \tan \delta \tag{5-4}$$

while at deeper layers

$$\mu = \tan \delta \tag{5-5}$$

(*LOTAR, 1979*; Fig. 5–25). The parameter μ^* in Equ. (5 – 4) is a parameter obtained from the grain size distribution curve of soils and can be expressed by

$$\mu^* = 1.2 + \log U \tag{5-6}$$

where U is the coefficient of uniformity (d_{60}/d_{10}).

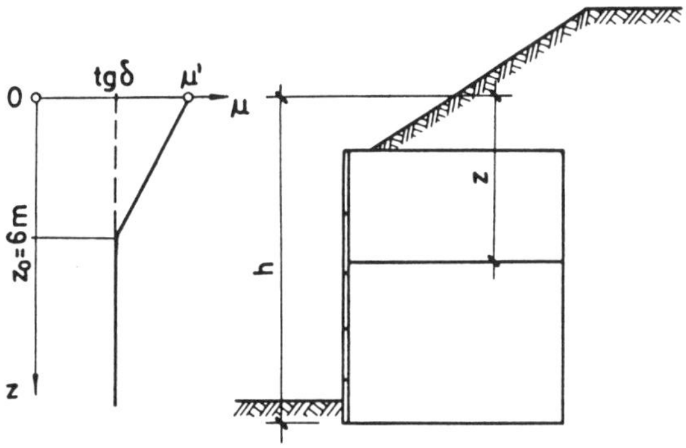

Figure 5–25 Internal frictional coefficient in the upper zone of a retaining wall

Frictional resistance can be increased significantly by using ribbed steel strips. The French design recommendations suggest a steel strip, for example, on which 3 mm high trapezoidal ribs are formed perpendicular to its longitudinal axis (Fig. 5–26).

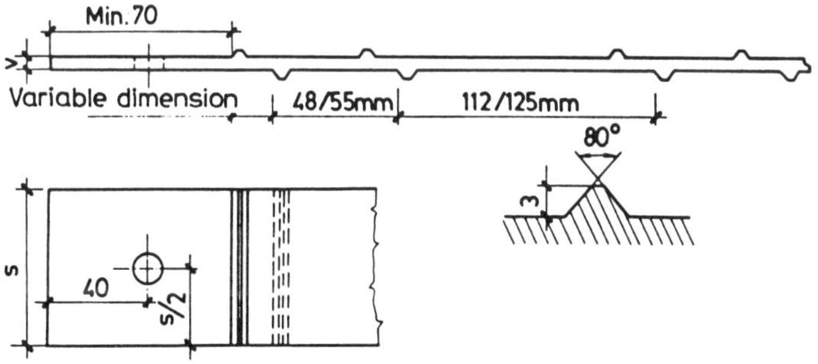

Figure 5–26 Ribbed strips

Similarly, in Hungary, favourable experience has been gained from shear tests performed in shear boxes with silty sand ($\varphi = 36 \ldots 37°$; $T_{r\rho} = 90 \ldots 95\%$, $w = 11\%$, $U = 7 \ldots 10$, $d_m = 0.05 \ldots 0.09$ mm; Table 5 4, *Szalatkay, 1983*).

Table 5–4 Frictional coefficient (μ) for clay and silty sand on smooth and ribbed steel plate

Characteristics	Soil type	
	Medium clay	Silty sand
Angle of internal friction	16°	37° –36°
Cohesion, kPa	470	20 –30
Frictional coefficient		
on smooth plate	0.1	0.4– 0.5
on ribbed plate	0.7	0.8– 0.6

5.2.3.2 Corrosion of metals in a soil environment

The durability of reinforcement is influenced by corrosion in soils. Steel strips become thinner and their cross-sectional area decreases so much that they may finally break. There are two kinds of corrosion attack: general (surface) and pitting corrosion. The latter occurs scattered over the surface area. The surface area of the strip also decreases as a consequence of general corrosion.

The rate of corrosion can be assumed to be constant and it can be expressed by a parameter with the dimensions μm/yr. It is influenced mostly by the type of soil and the material of the strip; its value can vary between 5 and 500 μm/year, i.e. between two orders of magnitude. Only few data are available concerning the corrosion of metal reinforcing elements and the deterioration of non-metallic reinforcement. Experience of 15...20 years is not sufficient to draw general conclusions. Corrosion rates are generally assessed cautiously, and, therefore, the amount of steel used in reinforced earth retaining walls, even with corrosion-protective coatings, usually exceeds the steel quantity which is strictly necessary.

The corrosion (and corrosion rate) are influenced by the

— electrochemical properties of the soil,
— types and conditions of soil and
— depth below the surface.

The list denotes the relative significance of the factors in decreasing order. Resistivity is considered to be most important parameter. If its value is small ($R < 700\ \Omega$cm), then the soil is classified as highly corrosive. If $R > 5000\ \Omega$cm, then the soil is noncorrosive. The measurement involves passing a known current through the soil and measuring the voltage drop along the line of the current flow; from these values the resistance is calculated.

Important factors influencing the soil aggressivity are the redox potential and the pH value. The former relates to the risk of microbiological corrosion and 'long-line corrosion' when a metal passes from an area of high oxygen content to an area of low oxygen content, and the latter represents the chemical behaviour in terms of acidity or alkalinity. If the pH value decreases (pH < 7), i.e. in acid soils, the corrosion hazard increases. The corrosion risk and corrosion rate can be estimated from the resistance and the pH value (*King, 1977*; Fig. 5–27). They can be measured by common laboratory methods.

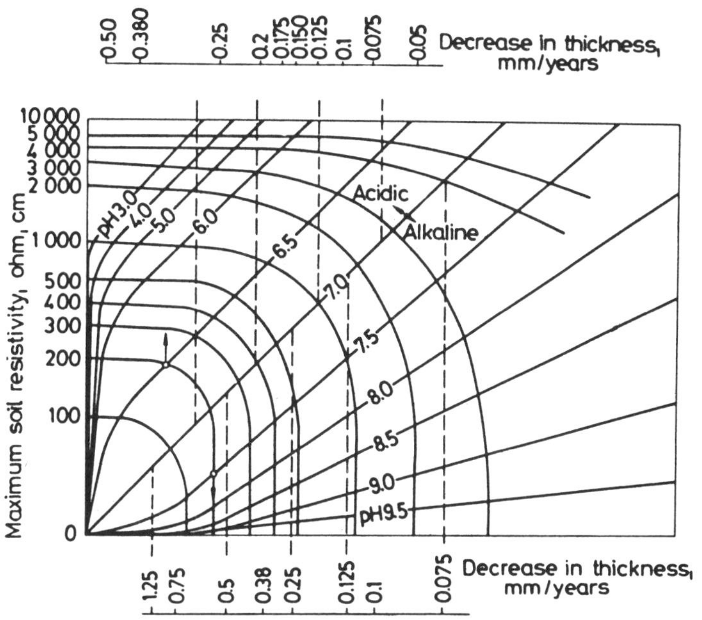

Figure 5–27 Determining the constant K_1 for the calculation of the corrosion rate (*King, 1977*)

The electrochemical properties of soils (and moisture of soil) are affected by their chemistry. As the salt—chloride, nitrate, sulphate—content increases, the conductivity (the reciprocal of the resistance) also increases. Organic materials can result in organic acids and metals buried in such soils being subjected to extensive pitting corrosion attacks. "The presence of organic material is also interrelated with hydrogen uptake. Hydrogen uptake rates in excess of 4.8 ml of hydrogen per 100 g of soil per day are taken as determining an aggressive soil. Published information indicates some correlation between soluble iron content and hydrogen uptake. Generally, the criterion accepted is that those soils, which contain more than 125 mg Fe/g soil are aggressive, while soils containing less than 53 mg Fe/g soil are not." (*King, 1977*).

The moisture content and rate of oxygen diffusion into the buried metal are both functions of the density and grading of the soil. Major corrosion damage occurs under conditions of alternate wetting and drying; thus corrosion at the shallow levels of the wall is more severe than at deeper ones. The strongest corrosion occurs in the upper 10...15 cm zone because of systematic drying and wetting. (It is of course quite likely that strips are not placed here.)

Aggressivity is at a maximum at about 20% moisture content, and general corrosion is most likely to occur. Below this value, the corrosion attack is more likely to be by pitting. The role of density, however, is contradictory: increasing the compaction decreases porosity and oxygen content but the conductivity of the denser soil mass is greater. From experience, however, piles driven into soils corrode less rapidly than those placed by drilling, so it appears that increasing the density is worthwhile. Corrosion hazards are less for granular soils with good particle distributions.

The dependence of corrosion on construction techniques can be related to the physical properties of soils, i.e. it is preferable to

— build in dry weather,
— compact effectively,
— drain the soil.

Table 5–5 Corrosion parameters for metal reinforcements (as defined in Code *TS–M57*)

Reinforcement	pH		Max. chloride content	Max. sulphate content	Min. specific resistance	Min. redox-potential	Parameters defining lifetime		
	min.	max.	%	%	Ω·cm	volt	K_1	K_2	n
Common structural steel	6	9	0.05	0.5	5000	0.4	see Fig. 5–27.	18	0.75
Aluminium alloys	6	8	0.05	0.5	3000	0.4	0.05	0	0.15
Stainless steel	5	10	0.03	0.5	5000	0.3	see Fig. 5–27.	0	0.1

A prediction of corrosion thickness (or depth) in time can be estimated from

$$v_k = K_1(\tau - K_2)^n \tag{5–7}$$

in which K_1, K_2 and n are constants; τ is the working life of the wall expressed in years. The constant K_1 depends on the electrochemical parameters of the soil and the strip, the values n and K_2 can be determined for the coating and material of the strip (Table 5–5, Fig. 5–27).

Protection against corrosion can be active and/or passive. Galvanizing with a zinc coating can give short term protection. Corrosion attacks can be decreased by selecting suitable soil and strip materials, and by applying construction methods which are appropriate to the chosen conditions. Protection is called passive if an additional thickness (corrosion allowance) is added to the nominal thickness.

5.2.3.3 Criteria for using soils and strips

In the last two paragraphs, the general criteria which guide the choice of the construction materials have been described; the various national design specifications are more or less agreed on these.

Adequate frictional resistance, compactibility and limited quantity of clay–silt content are the basic principles for selecting a suitable soil for the backfill. The first publications, which are still valid today, characterize those soils suitable for constructing embankments. On the basis of recent investigations and experience the specifications have been reviewed and, in some cases, relaxed.

According to the first French specification, the backfill shall contain not more than 15% material not exceeding the 75 μm particle size (*Vidal, 1969*). But according to recent specifications, it may contain 35% of particles not exceeding the 80 μm sieve size provided that one performs some appropriate physical tests. A complementary criterion specifies the minimum angle of internal friction

$$\varphi_{\min} = 25^\circ.$$

The *British Technical Memorandum* (Bridges)—Reinforced Earth Retaining Walls and Bridge Abutments for Embankments (BE 3/78) specifies a more rigorous criterion. Frictional as well as cohesive–frictional soil can be used for backfill of reinforced earth retaining walls.

Frictional fills shall comply with the following grading limits:

BS sieve size Percentage by mass passing	Grading limits
125 mm	100
90 mm	85–100
10 mm	25–100
600 μm	10–65
63 μm	0–10.

Cohesive frictional fill material passing the 425 μm BS sieve size shall have a liquid limit not exceeding 45% and a plasticity index not exceeding

20%. The fill shall contain not more than 10% material of less than 2 μm particle size and shall be finer than the following grading limits:

BS sieve size Percentage by mass passing	Grading limits
125 mm	100
90 mm	85
10 mm	25
600 μm	10
63 μm	10.

The German specifications for the minimum particle size are similar to those of the British or French: not more than 15% of all the particles by weight may be smaller than 60 μm. Not more than 25% of grains by weight may have dimensions greater than 100 mm (*VR—ABBE, Betonkalender II, 1978*). The angle of internal friction: $\varphi_{\min} = 25°$.

The Hungarian design specification (*TS–M–57*) for soil was compiled using the above considerations, and specifies that:

— the maximum allowed grain size is $d_{\max} = 250$ mm,

— not more than 25% by weight of grains should lie in the fraction of soil particles between the sizes $d = 150 \ldots 250$ mm,

— it should contain not more than 20% material not exceeding the 100 μm sieve size,

— it should contain not more than 10% material not exceeding the 20 μm sieve size,

— the coefficient of uniformity shall not be less than 7,

— the minimum internal angle of friction is $\varphi_{\min} = 25°$ (for the specified density).

The criteria detailed above seem to be very rigorous. In practice, it has been shown on the one hand that even with a greater quantity of silt and fine sand, the criterion of $\varphi_{\min} = 25°$ can be satisfied, and on the other hand two Hungarian retaining walls have proved that safe structures can be built using that type of soil (Fényeslitke, Siófok). There are some further cases where soils with a greater quantity of fine-graded material have been used. Measurements cited earlier and recent results of research prove possiblility of developing adhesion between clay and steel strips (*Potyondy, 1961*; *Ingold, 1981*).

The chemical properties of soils and reinforcement materials must be chosen so as to conform with each other. On the basis of British, French and German specifications it is suggested that:

— pH value should be between 5... 10,

— allowable minimum resistivity $R_{\min} = 5000\ \Omega$cm,

— maximum chloride ion content 500 mg/l,
— maximum sulphate ion content 5000 mg/l,
— maximum organic material content 2%.

These values are applicable for mild steel strips; for aluminium alloy and anticorrosive steel, constants and specifications are given in Table 5–5.

According to the *TS–M–57* standard cited earlier, mild steel strips have to be coated by hot-dip zinc galvanization to a thickness of 54 μm. A 54 μm thick coating gives safe protection against corrosion; thus the constant $K_2 = 18$ in Eq. (5–7) can be reduced. The corrosion allowance ν_k—the additional thickness of the strip—cannot be less than 1 mm.

There are no special criteria for processing the surface of the strip. The rougher the surface of the strip, the better is adhesion. The surface of plastic strips can be roughened by sand blasting. Ribs, too, improve the interaction between the two elements. It can be mentioned that nonuniformity in the soil deposits causes slight waviness along the strips which is not at all harmful; rather they improve the interaction to a certain degree. Crinkling and permanent deformation of strips, however, must be avoided since these can induce breakage later.

5.2.3.4 Facing elements and connections

Facing units for reinforced earth retaining walls are almost exclusively precast industrially or on construction sites. The form and appearance of reinforced precast concrete panels can be varied and they have to be designed with the following considerations:

— their dimensions and weight are determined by the lifting and fitting technology,
— neighbouring elements must be joined (by means of joints, holes, pins on the sides) to each other to provide temporary stability until the first soil layer is placed,
— joints must be designed to ensure easy assembly of the elements without breakage,
— by modifying the shape, grooving of the edges and finishing the surface of the concrete panels, it is possible to create any architectural form.

The thickness and reinforcement of the facing elements must be properly designed. At the points where strips are connected to the back of the facing elements, concentrated forces in equilibrium with the earth pressures have to be taken into consideration. In choosing the material for the facing elements, the influence of weather and corrosion must be accounted for (frost resistance, soil agressivity). The dimensions of the cruciform-shaped elements, applied originally in France, and well-proved in the Hungarian construction test (Fényeslitke, Siófok) are shown in

Fig. 5–28. Pins and holes on the shoulders and joints provide flexible interactions among the elements in the facing. The interlocking H-shaped elements shown in Fig. 5–19, also satisfy this requirement.

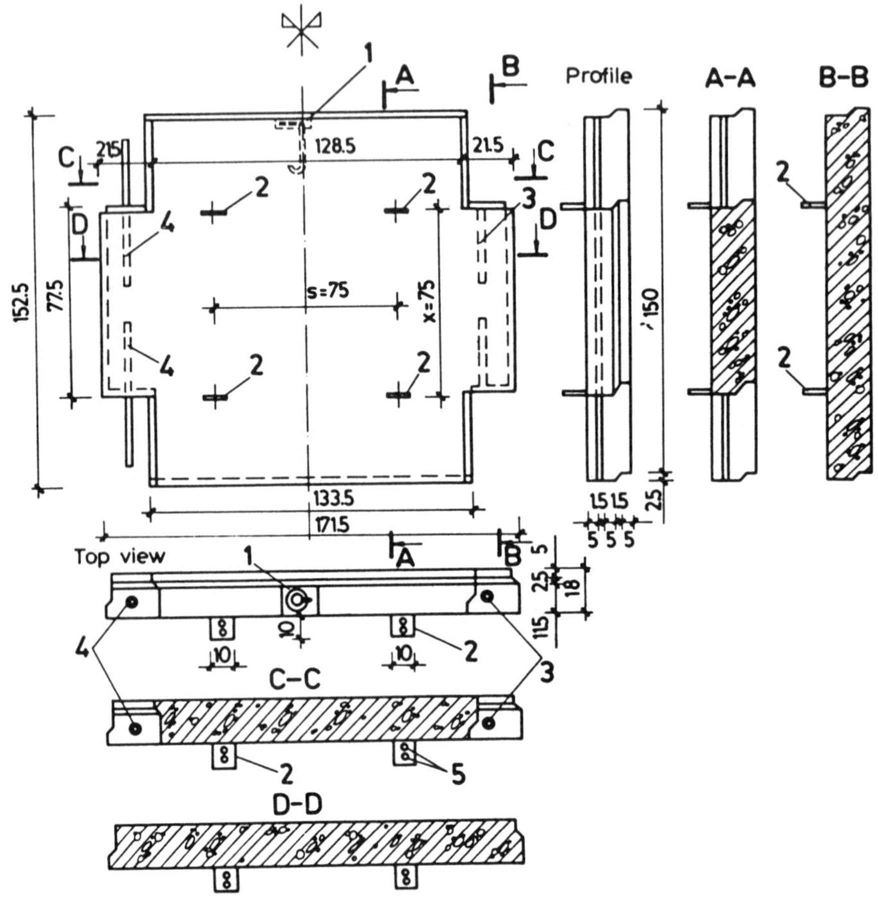

Figure 5–28 Precast reinforced concrete element used in the reinforced earth retaining wall built at Fényeslitke and Siófok

1 lifting hook; *2* ears to fix the strips; *3* plastic tube; *4* fitting rod; *5* hole for screw

Transferring the force from the strips to the strip-starters set into the facing elements is accomplished by means of connections (e.g. ears, bolts, rivets, mandrels, etc.). In the case of the York system, the hexagonal precast concrete wall units are threaded on poles to which the reinforcing elements—with the steel strips—are also attached (see Fig. 5–2). As for the connection system, the Hungarian H-type facing unit mentioned earlier is also interesting (Fig. 5–29).

Corrosion of the bolted connections in their special niches is prevented by subsequent concreting (*Marczal, 1981*).

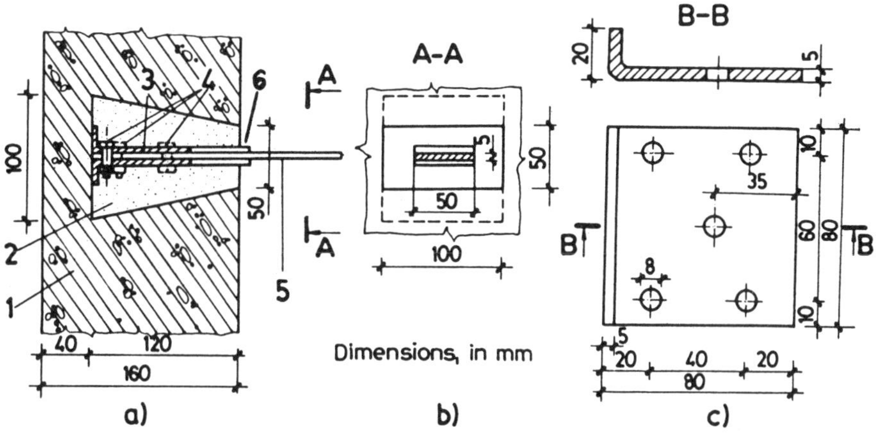

Figure 5–29 Fixing of the strip into the facing element, used by the Co. for Building Concrete Roads (*Marczal, 1981*)

a) clamping of the strip into the facing element (vertical cross section);
b) back view of the facing element; *c* fixing plate; *1* H-shaped reinforced concrete element;
2 anchoring seat (filled with Portland cement mortar); *3* pair of fixing plates (5 holes of ø 8 mm);
4 fixing bolts (M8 × 30); *5* reinforcing strip (50 × 5 mm); *6* rubber band inserts

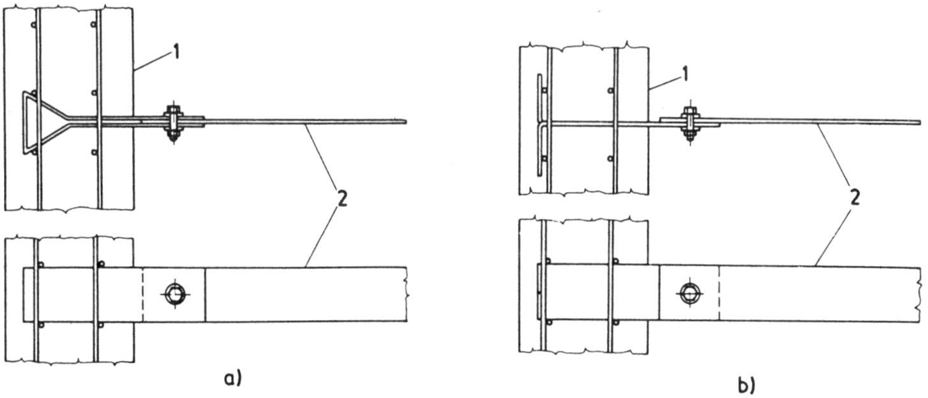

Figure 5–30 Ears for clamp

a) centric tension; *b)* eccentric tension; *1* facing element; *2* strip

The material of the strip starter is usually the same as that of the reinforcing strips. They have to be concreted into the element at the time of precasting, and therefore the strip spacings—the horizontal and vertical distances f and t between the strips—are fixed in advance. The maximum values of the strip forces are also known in advance, so that anchoring lengths, strip-starters and connection elements can be dimensioned. The connection to the strip-starter may be eccentric with respect to the strip (Fig. 5–30).

The appearance and shape of the complementary elements (closing, corner, half elements, etc.) should be in harmony with the basic elements. With the help of these structures, a wall of any height, not necessarily

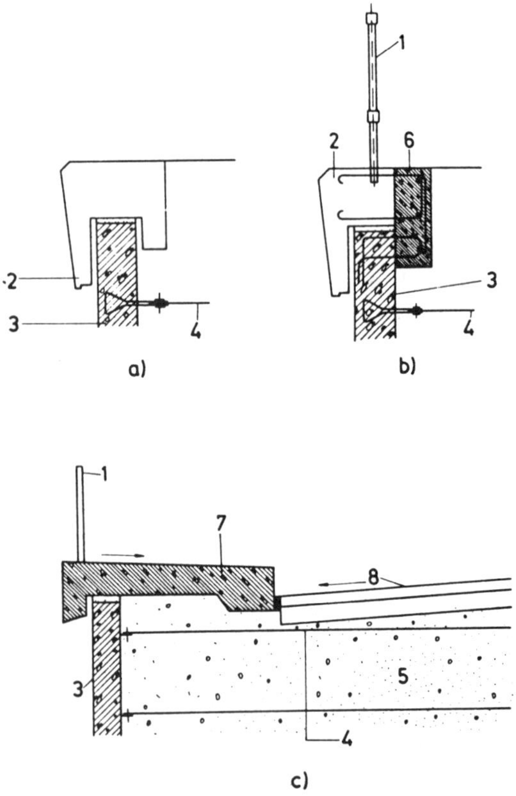

Figure 5–31 Capping for the facing

a) simple reinforced concrete capping;

b) capping assembled with vehicle parapet or guard rail, with monolithic concrete mass;

c) precast concrete pavement assembled with capping;

1 guard rail; *2* precast concrete element; *3* facing; *4* top row of strips; *5* fill;

6 monolithic concrete connection; *7* capping as precast concrete pavement; *8* pavement of the road equal to a multiple of the unit elements can be built. This is necessary whenever sloping embankments or backfills of various heights are to be constructed and the height of the facing must be gradually altered. Horizontal curves in a construction can be followed easily using 1.5 m wide elements.

On the surface of the reinforced construction precast reinforced capping, vehicle parapets, drainage channels, guard rails, side-walk pavements, bearing beams, etc. can be built. At the same time, the top edge of the wall must be finished aesthetically. Therefore, the elements listed above and the capping element should form a constructional unit—which may be precast (Fig. 5–31).

5.2.4 Designing reinforced earth retaining walls

Since the introduction of reinforced earth retaining walls, their analysis has developed a great deal. The objective of designers and research workers has been to form economic structures and to understand their mechanical behaviour and to develop theories to estimate their stability. Conventional design methods for retaining walls have frequently been applied until recently. The structural system has been studied using laboratory and large scale model tests as well as on-site investigations to develop theories for economic and reliable designs in conformity with their mechanical behaviour.

Mechanical analysis can be divided into two types according to two design conditions:

a) the ultimate limit state, and
b) the serviceability limit state.

The ultimate limit state corresponds to collapse or failure of a structure due to overstressing or rupture of the component elements. The serviceability limit state is that of excessive deformation of the reinforced earth retaining wall. This will be dependent on the required performance of the structure. In both analyses, the load-bearing capacity of all elements has to be established for all kinematically possible cases. The aim of a mechanical analysis for a reinforced retaining wall is to verify its environmental and external stability. There are methods for the analysis of reinforced earth retaining walls as well and the ideas and principles are presented in Chapter 4. Thus, here we do not deal with the external stability analyses.

The necessary conditions to achieve and maintain the interaction of the structural elements can be controlled using internal stability analyses. These methods have been developed recently and are developing further rapidly. In the following, they will be discussed in detail.

Serviceability limit analyses involve great difficulties. Inhomogeneity and anisotropy of the reinforced earth material are less disturbing since modern numerical methods and computer techniques make relatively accurate analysis possible. One unsolved problem at present, however, is that shortcomings in the various theories representing the mechanical behaviour of the soil cannot be eliminated by increasing numerical accuracy. Serviceability limit state analyses for reinforced retaining walls will be outlined in a later section.

5.2.4.1 Internal stability analysis of reinforced earth retaining walls

The fundamental question in the internal stability analysis is how to calculate the maximum tensile force in the strips and to determine the necessary dimensions and spacings of the strips. The calculations are based on the anchorage principle. The zone closest to the facing is called the active zone in which tangential stresses are directed towards the facing and the soil tends to pull out from the strips. The zone behind this is called the passive zone (or restraining zone) in which the tangential stress is directed inward, towards the fill and the soil tends to restrain the strips.

In the internal stability analysis, two criteria must be satisfied. The strips must be strong enough to prevent failure by breaking under tension, and sufficiently wide and long to prevent failure by pulling out.

According to the early design principles (*Schlosser, 1972*), the tensile force developed in the strips at depth h_i is (Fig. 5–32):

$$T_{ai} = \lambda_a \sigma_v f t \tag{5-8}$$

where σ_v = vertical earth pressure on the level of the strip. Symbols and notations here and in the following items are the same as in Section 5.1.

Assuming that the same stress σ_v acts on the restraining (passive) length l_p, the tensile force

$$T'_a = 2s\mu\sigma_v l_p \tag{5-9}$$

can be transmitted to the soil. In the extreme case $T_{ai} = T'_a$, from which it follows that

$$l_p = \frac{\lambda_a f t}{2\mu s} \tag{5-10}$$

therefore the restraining length of the applied strip must be greater than the value l_p.

To estimate the total length, the active length must also be known. Analyzing the tensile force distribution along the strips, *Schlosser* assumed that the maximum value developed on the boundary of the active

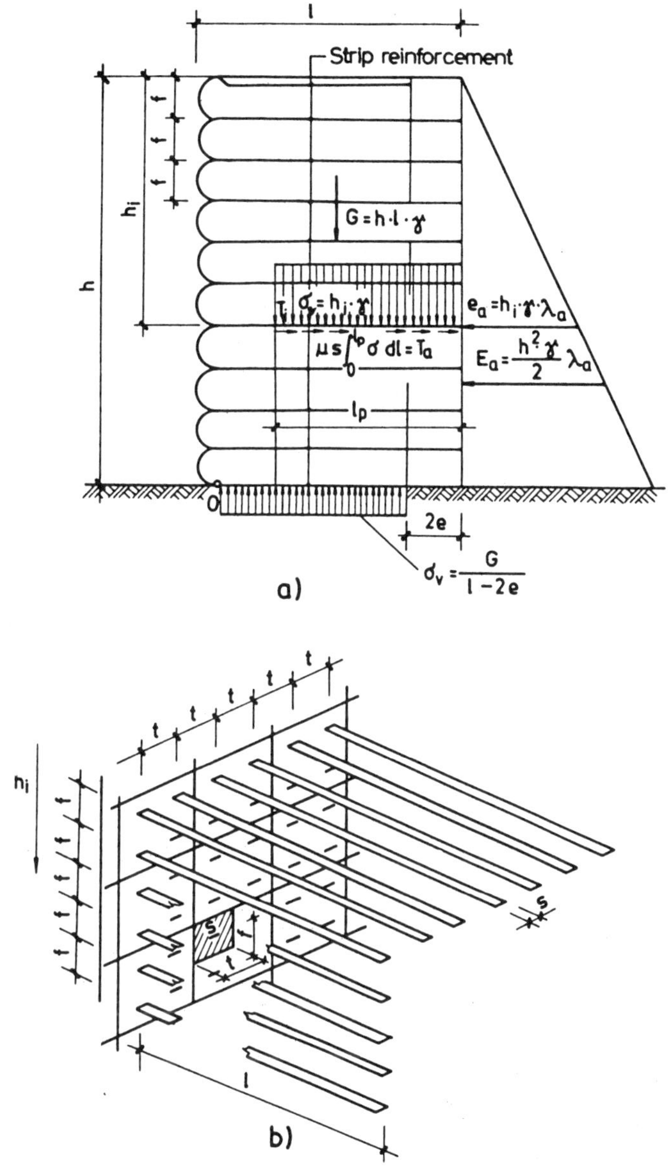

Figure 5–32 Internal stability calculation according to *Schlosser* (*1972*)

a) calculation of the earth pressure and the strip force;

b) main dimension of the reinforced earth retaining wall (2*e* eccentricity) and the passive lengths (Fig. 5–33); although originally he did not provide a theory to determine the exact location of this maximum.

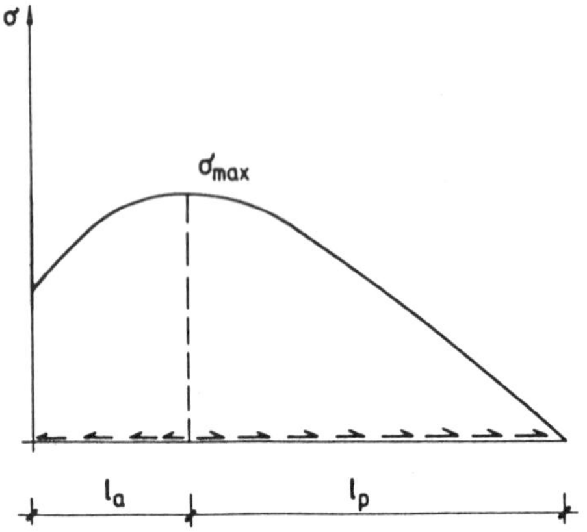

Figure 5–33 Distribution of tensile stress along a reinforcing strip
l_a active length of reinforcement; l_p passive length of reinforcement; σ tensile stress

For the total length he assumed

$$k = l/h = 0.8 \ldots 1.2 \tag{5–11}$$

the appropriateness of which has been verified in many practical cases.

Analysis of model and field tests conducted in the early 1970's, shows that the location of maximum tensile force along the strips describes the potential failure surface approximately. As an example, *Lee*'s theory (*1973*) is illustrative. This takes *Rankine*'s hypothesis into consideration in the calculation of earth pressure acting on the facing. He assumes that a failure plane at an angle of $\vartheta = 45° + \varphi/2$ to the horizontal develops inside the reinforced retaining wall (Fig. 5–34).

The wall is divided into two zones by this failure surface: the active one which slides downwards and the passive one which remains in its original place. Strips reach into the passive region and prevent the whole body from sliding. The strip force $T_{\mathrm{a\,max}}$ in the lowest level of the wall can be calculated by three methods:

— *Rankine* method:

$$T_{\mathrm{a\,max}} = h\gamma\lambda_{\mathrm{a}}tf \tag{5–12a}$$

— *Coulomb*'s force method

$$T_{\mathrm{a\,max}} = h\gamma\lambda_{\mathrm{a}}tf\frac{n}{n+1} \tag{5–12b}$$

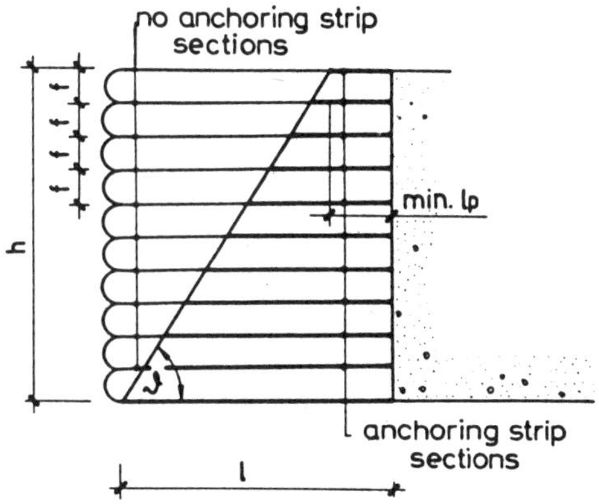

Figure 5–34 Internal stability calculation according to *Lee*

h height of retaining wall; *l* total length of the strip; *f* vertical spacing of the strips; l_a active length of strip; ϑ angle between the horizontal direction and the sliding surface

— *Coulomb*'s moment method

$$T_{a\max} = h\gamma\lambda_a t f \frac{n^2}{n^2-1}, \tag{5–12c}$$

where n is the number of the reinforcing layers. If $n \Rightarrow \infty$ the limiting values of the three forces are the same. The minimum length for the lowest strip is

$$l = \frac{h}{\tan\vartheta} + \frac{\lambda_a f t}{2s\mu}. \tag{5–13}$$

According to *Bacot*'s (*1975*) analysis the calculation of the displacement at failure is much more complicated. Two potential failure surfaces develop in a retaining wall having a height h and width l (Fig. 5–35); firstly the body ABCC′ and after that the body C′CB′ come apart from each other and from the backfill.

The failure plane, inclining at an angle of $\vartheta = 45° + \varphi/2$ to the horizontal starts from the point at the end of the strips—the first plane with the same angle and the second plane vertically (to a good approximation) reaches the surface of backfill. Equilibrium can be written for five forces:

— the weight of the trapezoid ABCC′:

$$G = \frac{h^2\gamma}{2}k(2 - k\tan\vartheta), \tag{5–14}$$

— the horizontal component of the earth pressure acting on the section backfill CC′:

$$E_{aH} = \frac{h^2\gamma}{2}\lambda_a(1 - k\tan\vartheta)^2, \tag{5–15}$$

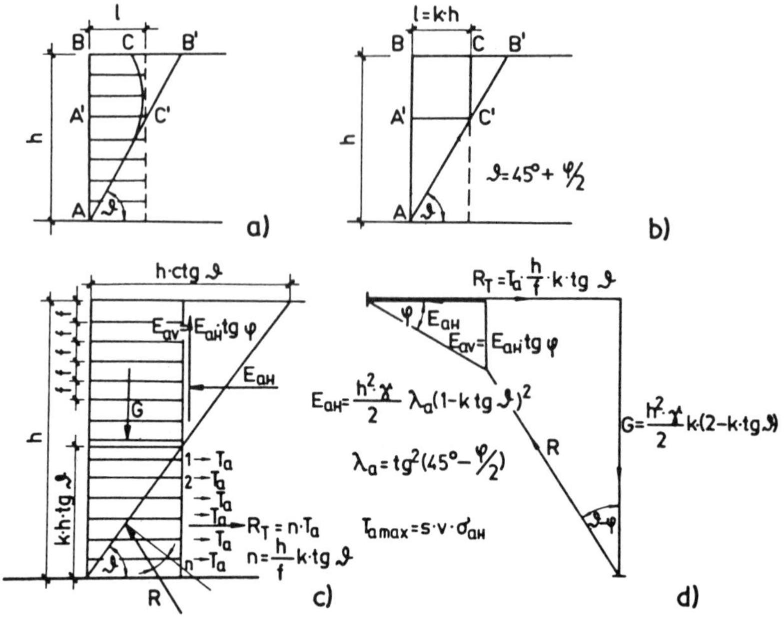

Figure 5–35 Internal stability calculation according to *Bacot* (*1975*)
a) development of the failure in the wall; *b)* double failure surface approach; *c)* forces; *d)* construction of the vectorial diagram; σ_{aH} allowable tensile strength for the steel strip

— the strip force acting on the plane AC′:

$$R_T = \frac{T_a}{f} kh \tan\vartheta, \qquad (5\text{–}16)$$

— the vertical component of the earth pressure acting on the section of the backfill CC′:

$$E_{aV} = E_{aH}\tan\varphi = \frac{h^2\gamma}{2}\lambda_a(1 - k\tan\vartheta)^2\tan\varphi, \qquad (5\text{–}17)$$

— the force acting on the line AC′:

$$R = \frac{G - E_{aH}\tan\varphi}{\cos(\vartheta - \varphi)}, \qquad (5\text{–}18)$$

where $k = l/h$.

The aim of the analysis is to obtain the critical height of the wall h_c, because the stability of a wall higher than this cannot be verified.

Therefore, a minimum criterion for h_c must be determined. Excluding the possibility that the strips slide out of the soil, by means of equilibrium at this level the relationship

$$\frac{T_a}{2skh^2\gamma(1 - k\tan\vartheta)} = \mu \tag{5–19}$$

is obtained. Rearranging this and expressing the limiting value criterion for the minimum

$$\frac{\partial h}{\partial k} = 0; \qquad k_{\min} = \frac{1}{2\tan\vartheta} \tag{5–20}$$

the following result is given:

$$h_{\min} = 2l\tan\vartheta. \tag{5–21}$$

With regard to construction: $h_{\min}$ denotes the maximum height of a stable reinforced earth retaining wall. For this critical height further estimates can be obtained elsewhere (*Kézdi, 1966*; *Smith, Bransby, 1976*; *Szalatkay, 1977, 1978*).

Experience in building practice, and the demand for economic improvements has focussed attention on the more complicated logarithmic spiral analysis which decreases the passive zone compared to the linear failure plane (*Juran, Schlosser, 1979*). For precast reinforced concrete facing elements, frictional resistance acting on the back of the facing can be considered in the calculation procedure. According to *Juran* and *Schlosser* the arching effect decreases the intercept D of the failure surface at the top level (Fig. 5–36); horizontal earth pressure acting on the facing increases linearly to the depth h_0 and below that it remains constant and the surface load acts down to depth h_0.

The calculation is based on the equilibrium criterion of the soil mass bounded by the facing and failure surface. Along the failure surface the validity of the *Kötter* relationship is assumed, according to which

$$\frac{\partial\sigma}{\partial l} + 2\sigma\tan\varphi\frac{\partial\varepsilon}{\partial l} = \gamma\cos(\varepsilon + \varphi)\cos\vartheta \tag{5–22}$$

where σ is the normal stress perpendicular to the sliding surface, ∂l is the elemental arc of the logarithmic spiral, ε is the angle between the vertical and the tangential direction at any point on the spiral.

Having solved the differential equation,

$$\sigma = \gamma r_0(\varepsilon_0, \varepsilon_f, \varepsilon, \varphi) + A_1$$

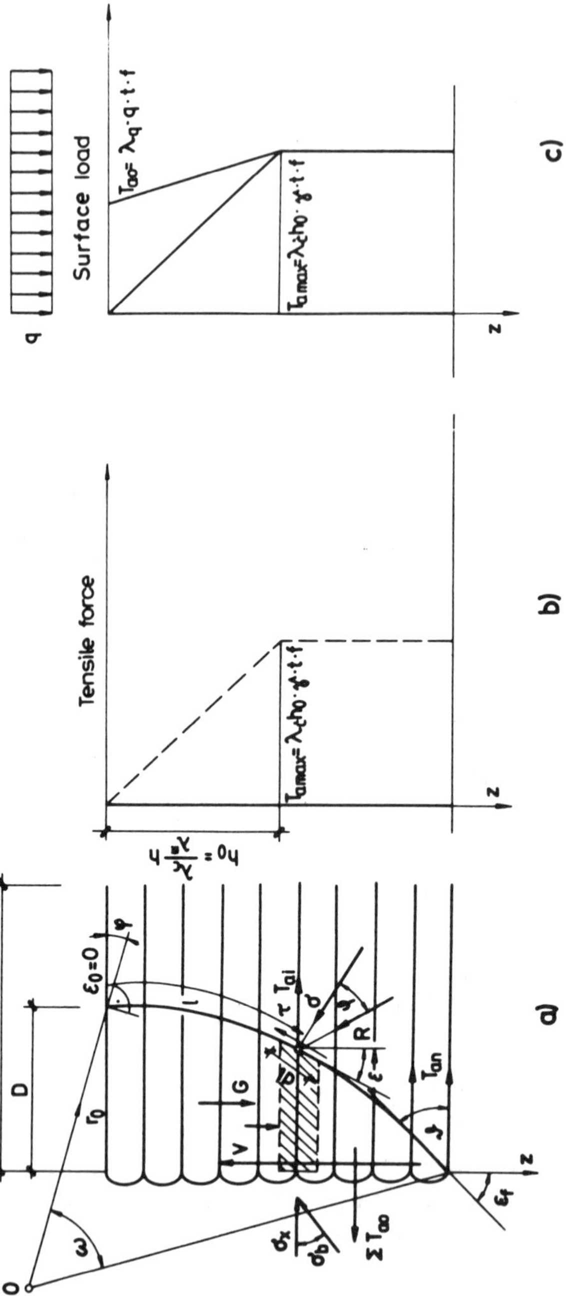

Figure 5–36 Internal stability analysis according to *Juran* and *Schlosser* (1979)

a) sliding surface and forces acting upon the active zone; *b)* calculation of the tensile force due to deadweight; *c)* calculation of the tensile force due to surface loads; σ_b angle of friction between soil and facing is obtained, where ε_0 and ε_f are the angles between the vertical and the tangents drawn to the start and end points of the failure surface; the constant A_1 can be calculated from the boundary condition $\sigma = 0$ corresponding to the boundary case $\varepsilon = \varepsilon_0$.

In the equation of equilibrium, there is the deadweight G of the sliding soil wedge, the horizontal resultant of the forces acting on the backfill, and the reaction force developing on the internal side of the facing.

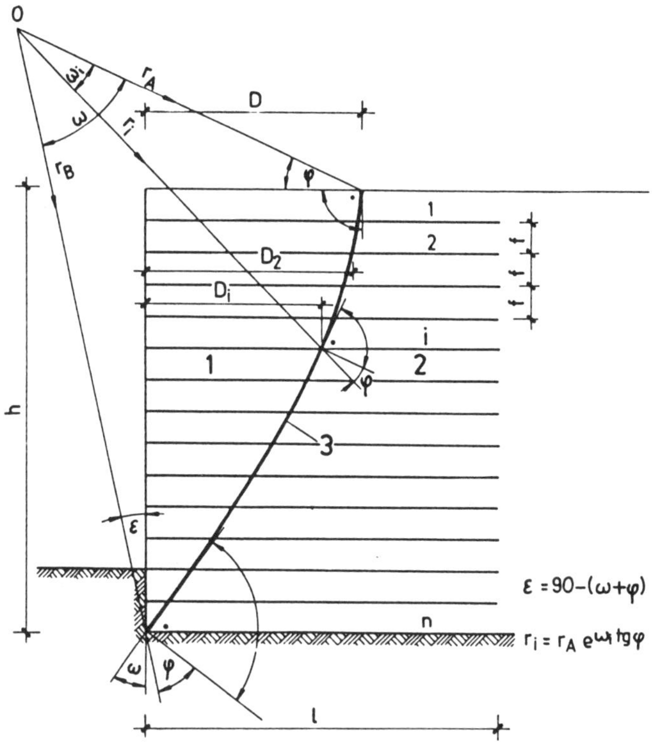

Figure 5–37 Construction of the logarithmic spiral sliding surface

1 active zone; *2* passive zone; *3* sliding surface

Having solved the static equations, the data necessary for constructing the logarithmic spiral as a function of the frictional angle can be obtained. As the Hungarian specification for design (TS–M–57) suggests the use of a logarithmic spiral failure surface, this procedure is described here.

The dimensions of the reinforced earth retaining wall are shown in Fig. 5–37; the constructional parameters described earlier are given in Table 5–6 (the parameters are determined from the angle of internal friction of the soil and the basic dimensions of the wall).

Table 5–6 Slide surface for different angles of soil friction

Angle of soil friction	Slide surface data		Earth pressure coefficient for max. tensile force	Earth pressure coefficient for surface load
φ	ω	D/h	λ_c	λ_q
25°	43.5°	0.43	0.26	0.70
30°	38°	0.37	0.21	0.60
35°	32.5°	0.31	0.18	0.50
40°	27°	0.26	0.15	0.41

Equation of the spiral is:

$$r_i = r_A e^{\omega_i \tan\varphi} \tag{5-23}$$

where r_A is the starting radius, ω_i is the central angle between an arbitrary radius r_i and the starting radius (Fig. 5–37). The analysis begins with the determination of the ratio D/h and the central angle ω (Table 5–6). When the angle $\varepsilon = 90° - (\omega + \varphi)$ and the intercept of failure surface D are known, the centre (0) and the length of the starting and the closing radii, r_A and r_B are obtained. In the knowledge of r_A and r_B, any radius length r_i can be calculated by means of the central angles ω_i (Fig. 5–37). Polar coordinates r_i, ω_i give any point on the spiral and the curve can be constructed by connecting the points with the desired accuracy.

Using the assumption that the maximum tensile force along the strips develops at the failure surface; its value is equal to the product of the horizontal earth pressure at the level of the strips and the surface area of the facing belonging to the strips, respectively.

In the case of an unloaded backfill surface, the horizontal earth pressure increases linearly to the boundary depth h_0; from there it is constant to the lowest level of the wall. Loads on the surface increase the value of the horizontal earth pressure but linearly decrease with depth down to the same depth h_0.

The steps of the calculation are as follows:

— the earth pressure coefficient:

$$\lambda^* = \frac{1 - \sin^2\varphi}{1 + \sin^2\varphi}, \tag{5-24}$$

— coefficient of the maximum strip force: λ_c,
— boundary depth h_0:

$$h_0 = \frac{\lambda_c}{\lambda^*} h, \qquad (5\text{–}25)$$

— the coefficient for the load acting on the surface of the backfill: λ_q (see Table 5-6),
— magnitude of the horizontal earth pressure on the surface of the backfill (Fig. 5-38) is the higher value of:

$$e_q = \lambda_q(q_a + 1.2q_e) \quad \text{or} \quad e_q = 1.2\lambda_q q_a \qquad (5\text{–}26)$$

where q_a is the sum of the dead loads and q_e is the sum of the live loads,
— intensity of the horizontal earth pressure at depth h_0 (Fig. 5-38):

$$e_{max} = \lambda^* h_0 \gamma_n \qquad (5\text{–}27a)$$

$$e_{max} = 1.2\lambda^* h_0 \gamma_n \qquad (5\text{–}27b)$$

where γ_n is the wet unit weight of the backfill after compaction.
— $e_1, e_2, \ldots e_i$ are the horizontal pressure ordinates at the levels $1, 2, \ldots i$, respectively.

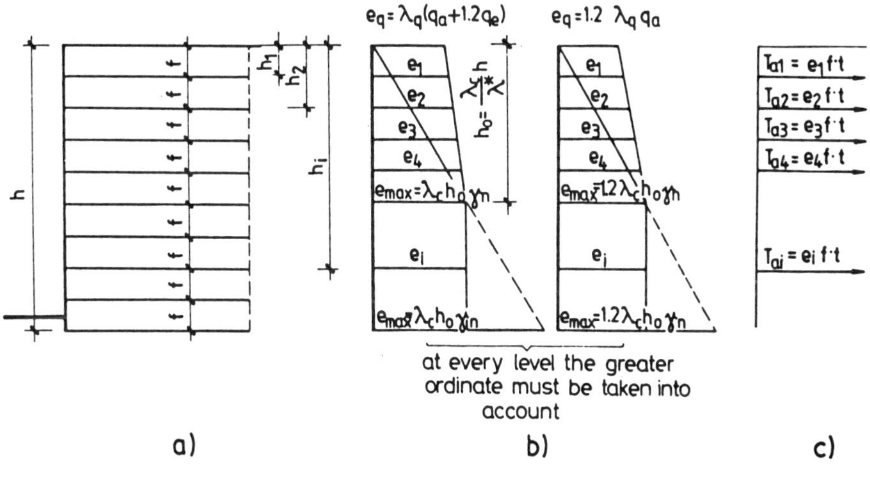

Figure 5-38 Calculation of the horizontal earth pressure and strip force
a) cross section of the wall; *b* earth pressure diagrams; *c* strip forces

Knowing the ordinates of the earth pressure, the strip forces are:

$$T_{ai} = e_i t f \qquad i = 1, \ldots, n \qquad (5\text{–}28)$$

(Fig. 5–38). The width of the strips is calculated at all levels using the criterion of frictional resistance:

$$s_i = n\frac{T_{ai}}{2(l - D_i)\mu(q_a + h_i\gamma_n)}, \tag{5–29}$$

where n is the safety factor, D_i is the length of the strip within the active zone. The required strip thickness is

$$v_{szi} = n\frac{T_{ai}}{s_i\sigma_H}, \tag{5–30}$$

where σ_H is the permissible stress for the strip. The suggested safety factor is 1.5.

5.2.4.2 Analysis of the serviceability limit state

In the preceding paragraphs, calculation methods based on the ultimate limit state were introduced: where failure is caused by a wedge-like earth mass breaking or displacing the strips. In these cases, safety against rupture is guaranteed by the safety factor. Consequently, the service state is far from the ultimate limit state of the wall, but we also need to calculate the deformations caused by these service loads. In the introduction to this chapter, we referred to the difficulties to be considered.

To determine the mechanical behaviour of walls under service loads, the most important tools of research and design practice are numerical methods based on the theories of continuum mechanics and the results of field investigations.

In accordance with the results of mechanics of composite materials, models of reinforced earth can be either 'discrete' or 'equivalent'. In the discrete model, the real dimensions and physical properties of the structural components are considered. One of the basic problems in the calculation method is to define the integration criterion appropriately. Mechanical properties for the equivalent model are determined separately and independently by analyzing the average behaviour of an individual strip and the surrounding soil.

Discrete models are usually more accurate than equivalent models, but they result in equations with many unknown variables and numerical problems. If relative displacement (sliding) is modelled at the interface between the soil and the strip, then the results indicate that greater deformations occur and the maximum of the strip force will move nearer the face (*Naylor, Richards, 1978*). When using finite element methods, (*Irons, Ahmad, 1979*; *Cookson, El-Zafrany, 1987*), there is no difficulty in separating the soil and the strip elements and to model the surface interaction by a special (string-like) element (Fig. 5–39*b*; *Andrawes, McGown, Mashour, Wilson-Fahmy, 1980*). For such cases, the soil and strip properties and some special constants must be determined by laboratory tests (Fig. 5–39).

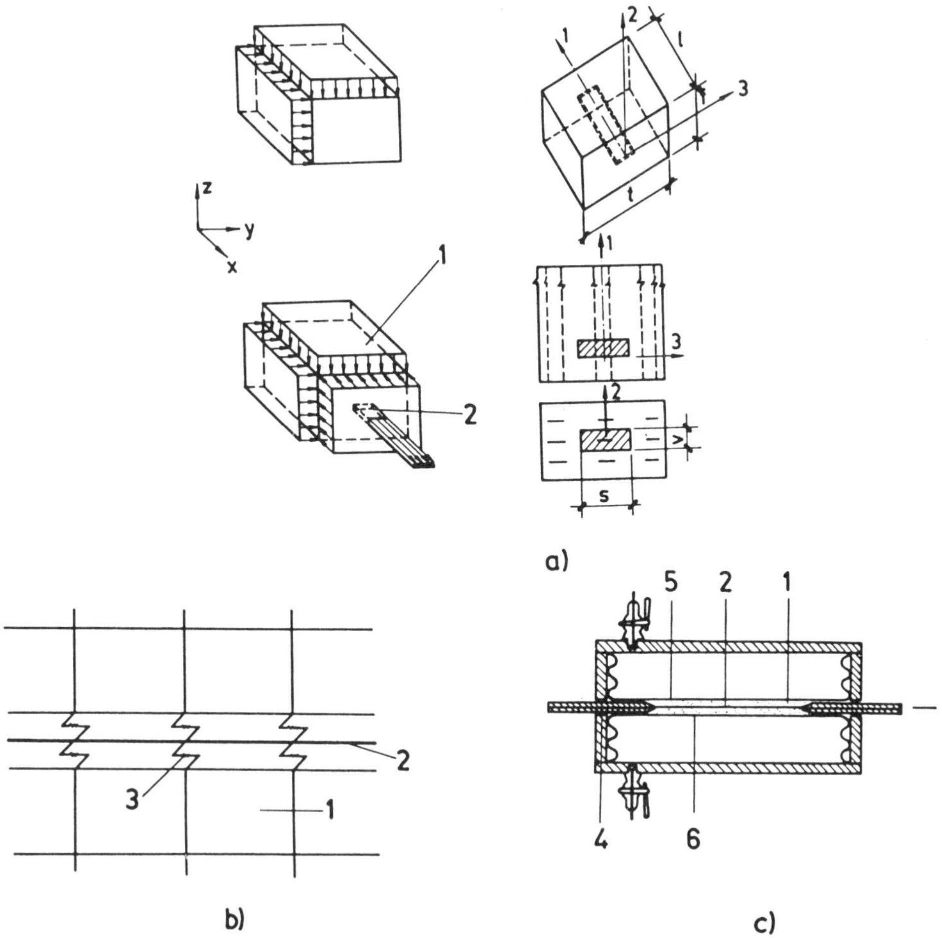

Figure 5–39 Mechanical model of the reinforced soil element

a) composite stresses (*Romstad, Herrmann, Shen, 1976*); *b)* discrete elements; *c)* experimental model (*Andrawes, McGown, Mashour, Wilson-Fahmy, 1980*); *1* soil; *2* inclusion; *3* spring element; *4* sliding without friction; *5* water; *6* rubber membrane

Methods using the 'equivalent material properties' are numerically simpler. Reinforced earth can be regarded as a transverse anisotropic or orthotropic material; parameters belonging to the equivalent continuum must be estimated by mechanical analyses (Fig. 5–39; *Romstad,*

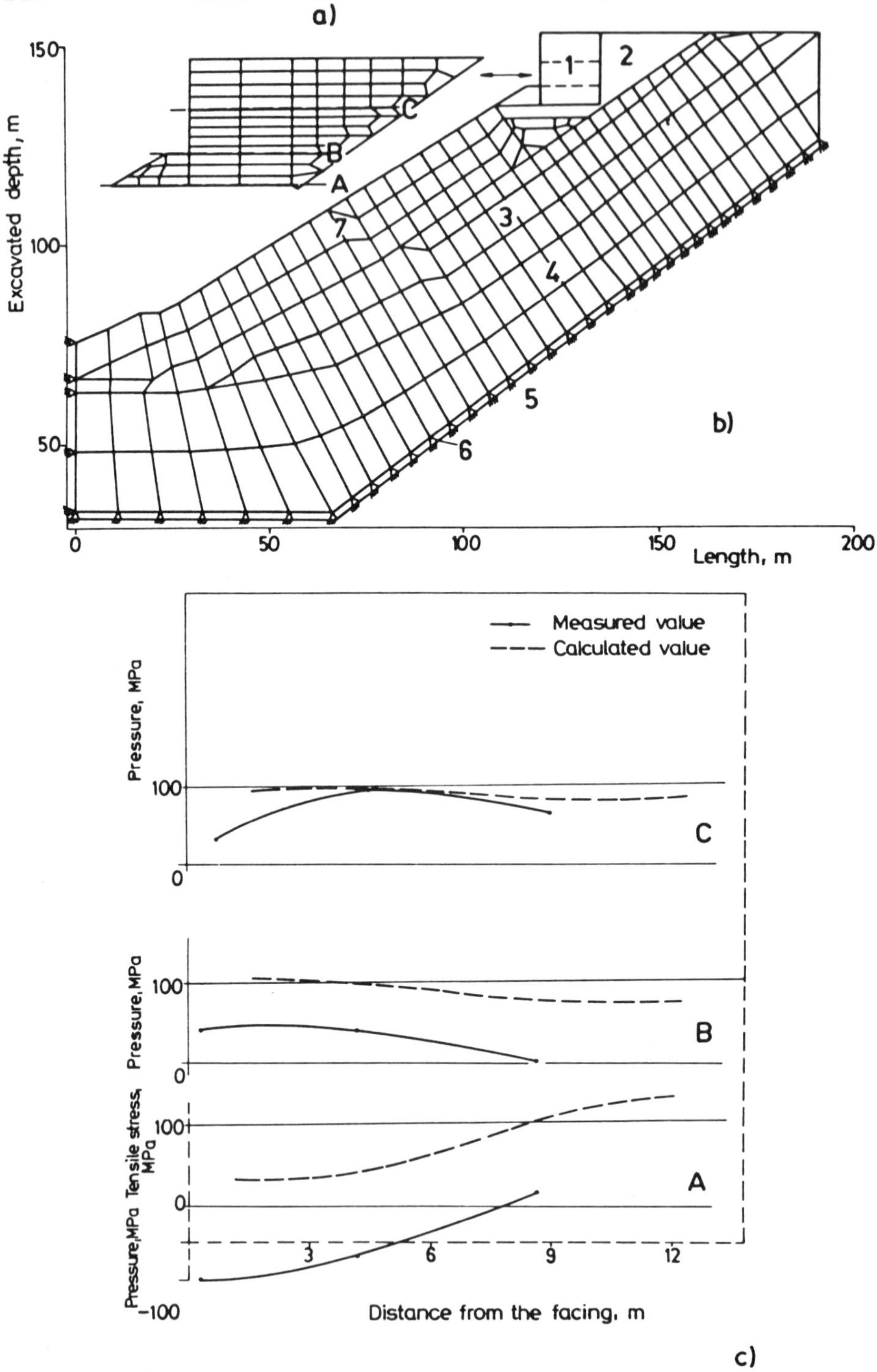

Figure 5–40 Finite element mesh (*Chang, Forsyth, 1977*)

a) retaining wall; *b)* surrounding of the wall; *c)* comparison of results of measurement and calculation; *1* retaining wall; *2* backfill; *3* potential failure surface; *4* original fill; *5* base rock; *6* fixed connections; *7* fill; *A*, *B*, *C* measured and calculated levels

Herrmann, Shen, 1976). In the case of a simple transversely anisotropic model, five material parameters are needed. The orthotropic model must be applied when the reinforcing layer is anisotropic in its own plane. The classic types of reinforced earth retaining walls are of that kind. This does not cause problems in usual calculations because the stress state in the section parallel to the facing is not analysed in general. For the determination of the material properties in the equivalent continuum models, the physical nonlinearity or certain singularities (sliding, yielding, etc.) can also be considered (*Herrmann, Al-Yassin, 1978*). These methods are based on the composite stress principle (in which equivalent strains and stresses are compared) or on the equivalency of strain energies (*Harrison, Gerrard, 1972*; *Gerrard, 1982*).

In Fig. 5–40, a finite element mesh for a retaining wall is shown. It reaches far beyond the region of the wall, and the boundary conditions correspond to a rock surface below the wall, where no displacement occurs. The free surface at the top of the wall can model cracking.

The results of field measurements and the calculations of the strip forces at three levels are shown in the diagrams that follow. The correlation is not completely convincing; the differences prove that the model is, in some cases, very good but in others it gives unacceptable results.

The differences between the calculated and measured values do not prove that the calculations are imperfect by any means; they may originate from poor quality measurements.

Nonetheless, the reliability of measurements is usually satisfactory and, in general, they are very useful for developing theoretical models.

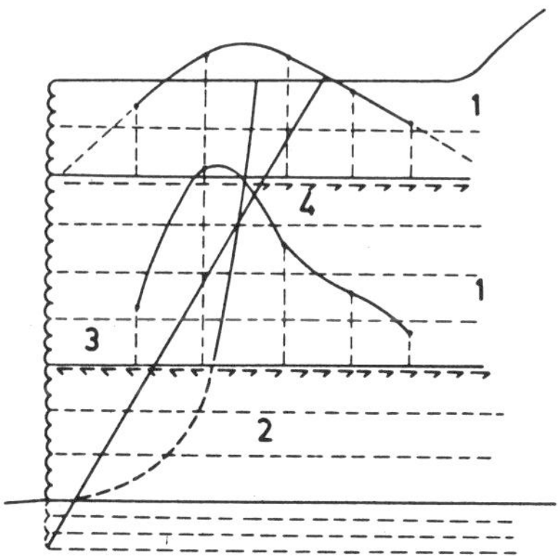

Figure 5–41 Results of measurements of strip forces (*Schlosser, 1972*)

1 distribution of the strip force; *2* curve connecting the maxima of the tensile forces; *3* active zone; *4* passive zone

For example, on the basis of the first strip force measurements (*Schlosser, 1972*) an early hypothesis regarding the location of the maximum strip force was disproved (Fig. 5–41).

Since then, strip forces have been measured by many workers, (*Sohnius, 1982*); straingauges with waterproof and protective coatings are usually used. Deformation measurements of the facing are considered to be the simplest, because only ordinary surveying instruments are necessary to perform this task. Carrying out these measurements is justified in every case, because by this means any unusual behaviour can be registered immediately (e.g. large displacement caused by overcompaction of the backfill). The results of measurements during a test loading are presented in Fig. 5–42 (*Al-Hussaini, Perry, 1978*). From experience to date, deformations of the facing of well-designed and well-constructed reinforced earth retaining walls are usually similar to that shown in this figure.

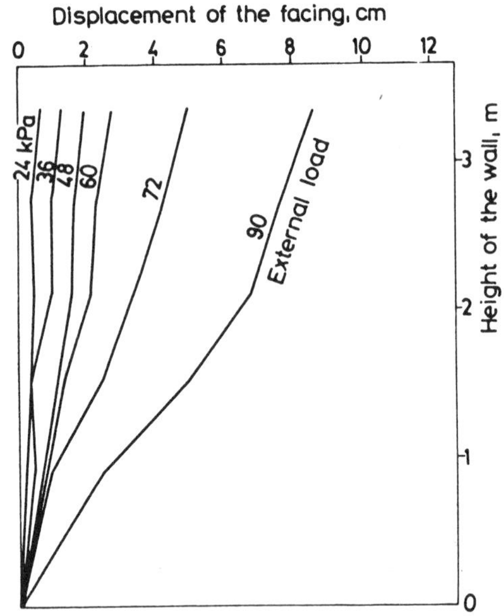

Figure 5–42 Deformation of facing due to external load

Some simple field investigations were performed also during the Hungarian experimental work. The results proved that soil and construction conditions met the usual expectations (*Magyarné, Radnóti, Scharle, Szalatkay, 1979*; *Szalatkay, 1981*; *Marczal, 1982*).

5.2.5 Design of a reinforced earth retaining wall

In a narrow sense, the steps of design are as follows:
- preliminary design,
- stability analysis,
- design of the facing elements,
- design of the complementary elements,
- design of the connections between the facing elements and the strips,
- design of drainage.

In the preliminary work, soil conditions must be explored and the stratification and properties of the soils must be determined in order to perform the appropriate external stability analyses. The backfill soil materials must be investigated from both the physical and the chemical points of view. Grading, compactibility and angle of internal friction of the soil should be determined in the laboratory. The pH value, electrical resistance, redox potential, organic material content, sulphate and chloride ion concentration should be measured. From these results, the suitability of the soil as backfill can be assessed and the appropriate strip material can be selected. These properties are then used in the internal stability analyses as well as in the corrosion calculation.

Having determined the width and foundation depth of the wall, the internal stability analyses can be performed. Loads acting on the wall can be divided into dead loads and live loads. Dead loads to be considered are as follows: (q_a):
- deadweight of the fill,
- uniformly distributed surcharge,
- foundation loads from bridges,
- self-weight of the permanent structures (pavement, sub-base, etc.) on the backfill,
- earth pressures developing from the dead loads.

Occasional loads might be (q_e):
- vehicle and pedestrian loads,
- temporary loads during construction,
- earth pressures resulting from live loads.

For the calculations, the frictional coefficient between the soil and the strip should be measured in the laboratory if possible. If this is not possible, then $\mu = 0.4$ may be assumed (smooth steel strip). The necessary thickness (v_{sz}) is obtained from Eq. 5–30. Having calculated the corrosion allowance (v_k) the total thickness of the strip is:

$$v = v_{sz} + v_k$$

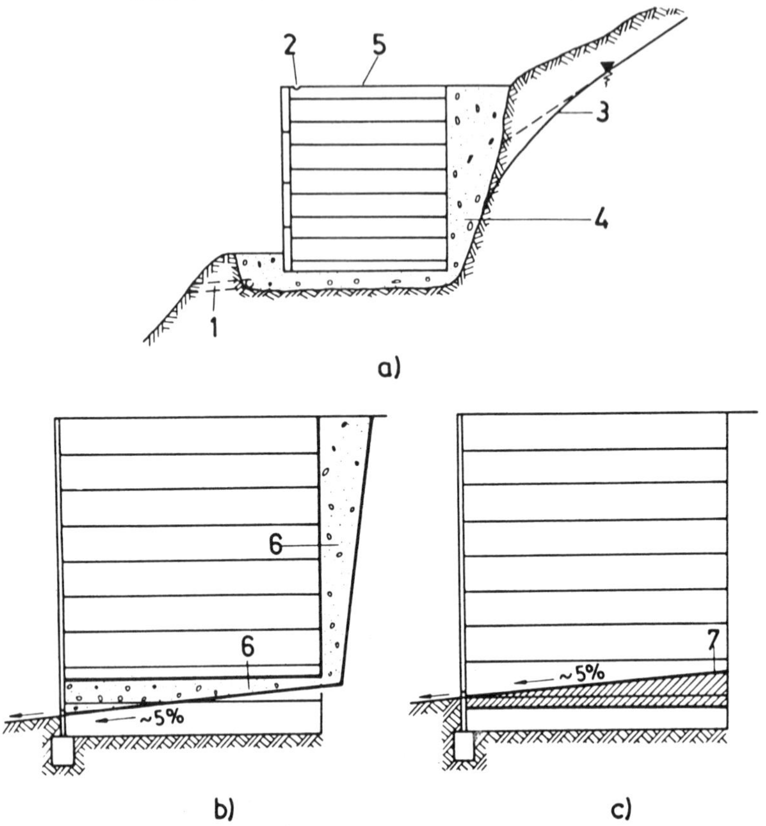

Figure 5–43 Drainage conditions

a) water in the soil and sloped surface; *b)* less permeable fill; *c)* permeable fill;
1 drain; *2* drainage channel; *3* water-table depression due to the draining layer;
4 groundwater conductive layer; *5* retaining wall; *6* horizontal drain; *7* stabilized, impervious layer

where v_k is not less then 1 mm.

The technical specifications of the facing elements (as well as complementary ones) relate to their weight, their connections, aesthetic appearance and dimensional tolerance. Connecting elements between strip and facing must be designed for the lowest strip force and the bolt connections must be checked against shearing.

Particular attention must be paid to drainage in the case of reinforced earth walls. By removing water, the dangers of instability and corrosion are decreased. Changes in water level can be especially dangerous.

Drainage design is influenced by site conditions and the purpose of the structure.

Usually the
— morphology,
— subsoil-water conditions,
— permeability and aggressivity for concrete of the soil
must be considered. The backfills used for reinforced earth retaining walls have good frictional characteristics, which usually means that they also have good drainage properties. Special precautions should be taken in cases where the backfill does not have good drainage. In addition, drains, impervious layers, drainage paths, topsoil and pavements on the surface of backfill all influence the drainage conditions.

General rules for drainage cannot be given but the examples in Figs 5–43 and 5–44 may be helpful in practice.

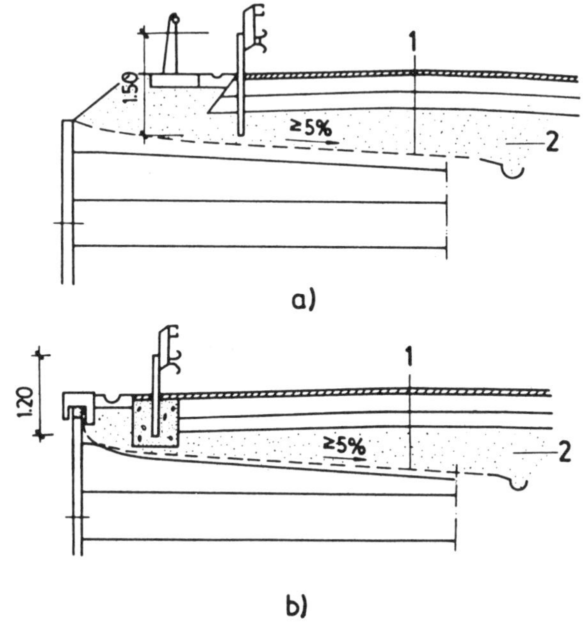

Figure 5–44 Drainage layer below a pavement in a reinforced earth retaining wall
a) sloped backfill; *b)* smooth backfill; *1* geotextile; *2* drain

5.2.6 Construction of reinforced earth retaining walls

The building of reinforced earth retaining walls consists of a series of successive stages:
— positioning the facing elements,
— deposition and compaction of the earth fill,
— assembly of the reinforcing strips.

In most cases, the concrete facing elements are lifted and transported by a light crane on the free side of the wall. For this purpose, accessibility in front of the facing must be provided. The most favourable condition is when the crane is able to pass up and down parallel to the line of the facing. It is usually practical to store the facing elements on the opposite side of the passage in front of the wall. From here with only a 180° turn, a crane can place the elements into their positions. The alignment must be controlled by surveying instruments, and to keep them in their correct positions, wooden wedges and temporary clamps may be used. To ensure stability when the first layer of the elements is positioned, temporary supports are installed on the facing side of the wall, which are then removed as soon as the backfill is placed.

Assembling reinforced strips and depositing, spreading, levelling and compacting the soil backfill are critical operations in the erection sequence. The reinforcement strips are placed perpendicular to the wall face on the previously placed layer of compacted earth fill. "Deposition, spreading, levelling and compaction of the fill shall be carried out generally in a direction parallel to the facing and shall be executed in stages to alternate with the placing and fixing of the reinforcing elements and the facing. Care should be taken to ensure that reinforcing elements and facing are not damaged or displaced during deposition, spreading, levelling and compaction of the fill. The programme of filling shall be arranged so that no machines or vehicles should run on the reinforcing elements" (RE–TM). It is advisable to restrict earth moving equipment to 2 m from the face of the wall to avoid any shifting of the facing elements which have not yet been fully backfilled. It is imperative that the trucks should not come in direct contact with the reinforcing strips. The trucks should pass only over layers of earth fill (*Legrand, Schlosser, 1976*).

It is suggested that the fill should be compacted to 90...95% of the maximum dry density. Overcompaction can cause deleterious deformations of the facing. It is also suggested that the optimal compaction method and number of passes should be tested for the given soil conditions (optimum moisture content, type of the wall, thickness of soil layers, etc.) and compaction machine.

During construction inspection, it is necessary to ensure the proper building quality. Apart from the usual means of quality control associated with road building, special care must be taken to measure the

— density of the fill,
— deformation of the facing.

The latter operation is needed to avoid overcompaction of the backfill. Allowances for density variations and allowable maximum values in tilting or sliding of the face must be specified by the designer. Aesthetically acceptable tilting is about $\Delta = 1/100 \ldots 1/200$.

The finishing works depend on the main purpose of the reinforced earth structure:

— placing precast or cast-in-situ capping reinforçing elements on the backfill,
— placing topsoil and fill on the wall,
— placing fences, pavements and parapets along the edges of the wall.

5.3 REINFORCING BY SHEETS

In the introduction of this chapter it was mentioned that sheets (grids, nets, geotextiles, etc.) are increasingly being used for reinforced earth retaining walls. These materials, appropriately for reinforcement (Table 5–1), are more flexible but have significantly smaller strengths compared with metals. However, as they cover the total surface of the whole soil mass, they can bear the earth pressure of 7...8 m high retaining walls, if the reinforcement is placed at 0.3...0.5 m spacings. They must be laid more densely (by compacting the soil in thinner layers) generally and have greater frictional resistance than metal strips. The behaviour of retaining walls reinforced by sheets is close to that of gravity walls and is governed by the principle of block-like support.

The principle was elaborated long ago (Fig. 5–3). At the first conference on geotextiles held in Paris, many authors introduced similar construction systems and many of them suggested design methods analogous to those for reinforced earth retaining walls (*Bell, Steward, 1977*; *Holtz, Broms, 1977*; *Kern, 1977*; *Puig, Blivet, Pasquet, 1977*). Specialists mostly think it is an advantage to use flexible sheets, although this is not always the case (abutments for example are sensitive to differential settlement).

In addition to the relatively high deformations, lack of knowledge relating to long-term strength and fatigue of plastics make users cautious. Because of imperfect knowledge of the effects of soil corrosion and climatic effects (frost, ultra-violet rays), reinforcement with nets are recommended only for smaller temporary constructions designed for short service lives.

From cost analyses, retaining walls reinforced by sheets or fabrics are very cheap (*Risse, Göbel, Hanspach, Piesker, 1982*). Assuming the same efficiency, a reinforced earth retaining wall is by 20...30%, the gravity wall 70...100%, and steel sheet pile 120...140% more expensive than walls reinforced by geotextiles (Fig. 5–45).

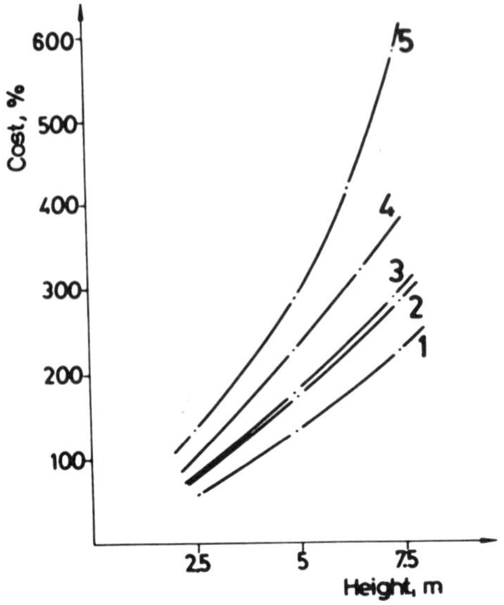

Figure 5–45 Comparative cost analysis of various retaining walls (*Hanspach, Piesker, Risse, Göbel, 1982*)
basis: 2.5 m high gravity wall = 100%
1 earth retaining wall reinforced by sheets;
2 earth retaining wall with precast reinforced concrete panels;
3 earth retaining wall with steel facing; *4* gravity wall; *5* steel sheet-piles

Estimates of service life are shown by the data presented in Table 5–7 (*Piesker, Göbel, 1982*). Without facing, only two years life and with steel or asbestos (cement) sheeting at most twenty years service life can be expected.

In the following some typical examples and designs of earth retaining walls reinforced by sheets are presented without aiming at completeness. Further examples are given in Chapter 7.

5.3.1 Construction methods and examples of earth walls reinforced by sheets

According to the construction methods two kinds of wall can be distinguished:

- facing is formed by bending the reinforcement sheets over backwards,
- facing is separated from the reinforcement and formed by using different constructional elements.

Table 5–7 Structural arrangement of soil walls reinforced with geotextiles (*Piesker, Göbel,* 1982)

Designed lifetime	Arrangement principle	Type of material* *a)* as reinforcement *b)* as covering/facing	Type of plastics**
2 years	1. 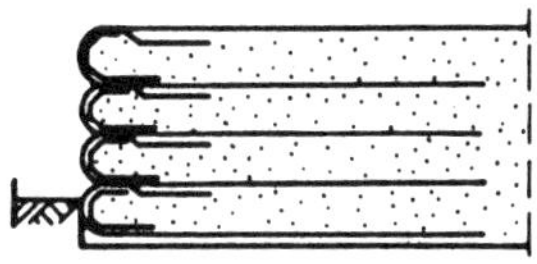 No covering on face	*a)* TPE 170	PE
10 years	2.1 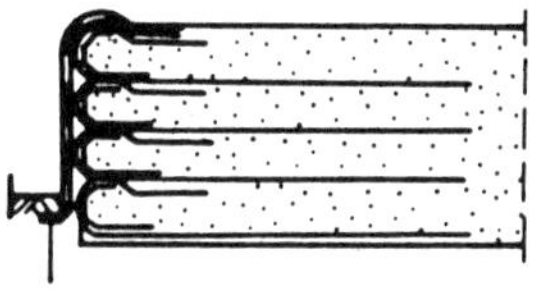 Geotextile facing	*a)* TPA/PE 250 TPA/PE 360 *b)* SPE 170 SPA 250; SPA 360	PA/PE PA/PE PE PA/PE
	2.2 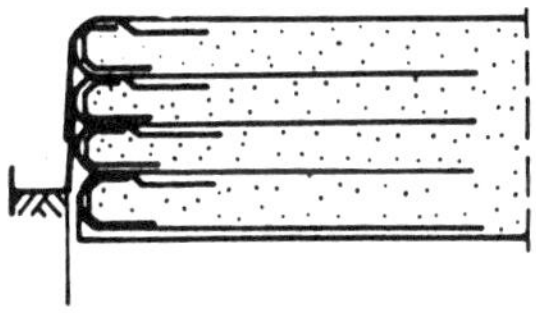 Facing with special plate elements	*a)* TPA/PE 250 TPA/PE 360 *b)* Corrugated asbestos cement sheet	*a)* PA/PE PA/PE
	2.3 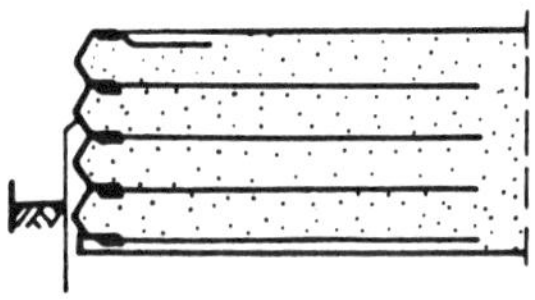 Facing with steel elements	*a)* TPA/PE 250 TPA/PE 360	PA/PE PA/PE —
20 years	3.1 as 2.1	*a)* TPE 170 *b)* SPE 170, SPA 250	PE, PE, PA/PE
	3.2 as 2.2	*a)* TPE 190 *b)* as 2.2*b)*	PE —
	3.3 as 2.3	*a)* TPE 170 *b)* —	PE

* Notation used in Germany
** PE = polyester; PA = polyamide

The two solutions hardly differ from each other from the mechanical point of view. In Fig. 5–46, four variations can be seen: the first two illustrate the separate system, the other two illustrate the other method.

Both types of construction have advantages and disadvantages. Problems may occur at the connections if the reinforcement is separated from the facing and, in the other case, alignment can cause difficulties during construction.

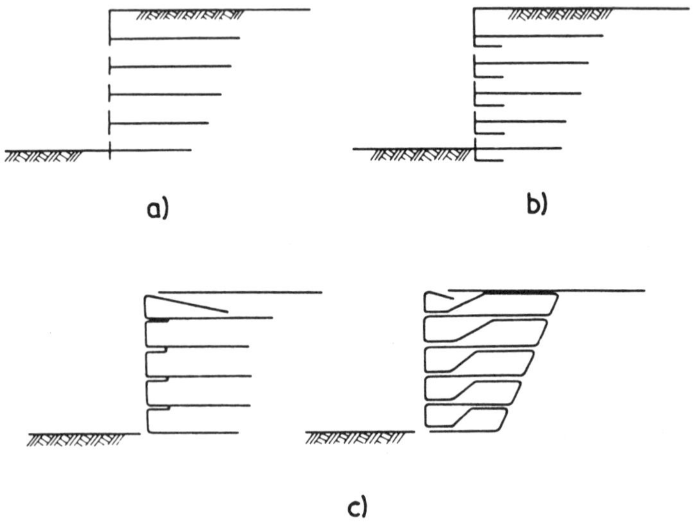

Figure 5–46 Some actual and proposed wall element-, and anchor geometries for reinforced earth retaining wall

a) strips attached to flat wall element (*Al-Hussaini, 1977*);

b) strips attached to folded wall element (*Broms, 1977*);

c) layering anchor sheets with soil to form the wall face (*Bell and Steward, 1977*; *Nicholls, 1981*)

A remarkable example is shown in Fig. 5–47. At the Newmarket Silkstone Colliery near Wakefield, (North Yorks, in Britain) a retaining wall was built to support a 40-ton locomotive and eight 25-ton trucks.

A precast prestressed double unit, 2.4 m wide, was used and for reinforcing an ICI Fibretain glass fibre was used, 5 m long, fastened to the structural elements by bolts. Minestone waste from the colliery was used as backfill and Netlon Tensar, SR high tensile polymer grids fixed by helix connections were used as reinforcing elements. The width of the reinforcement was 1.0 m, and the distance between their centres was 1.2 m; thus they almost completely covered the horizontal section of the backfill. Vertical spacing between them was 25 cm.

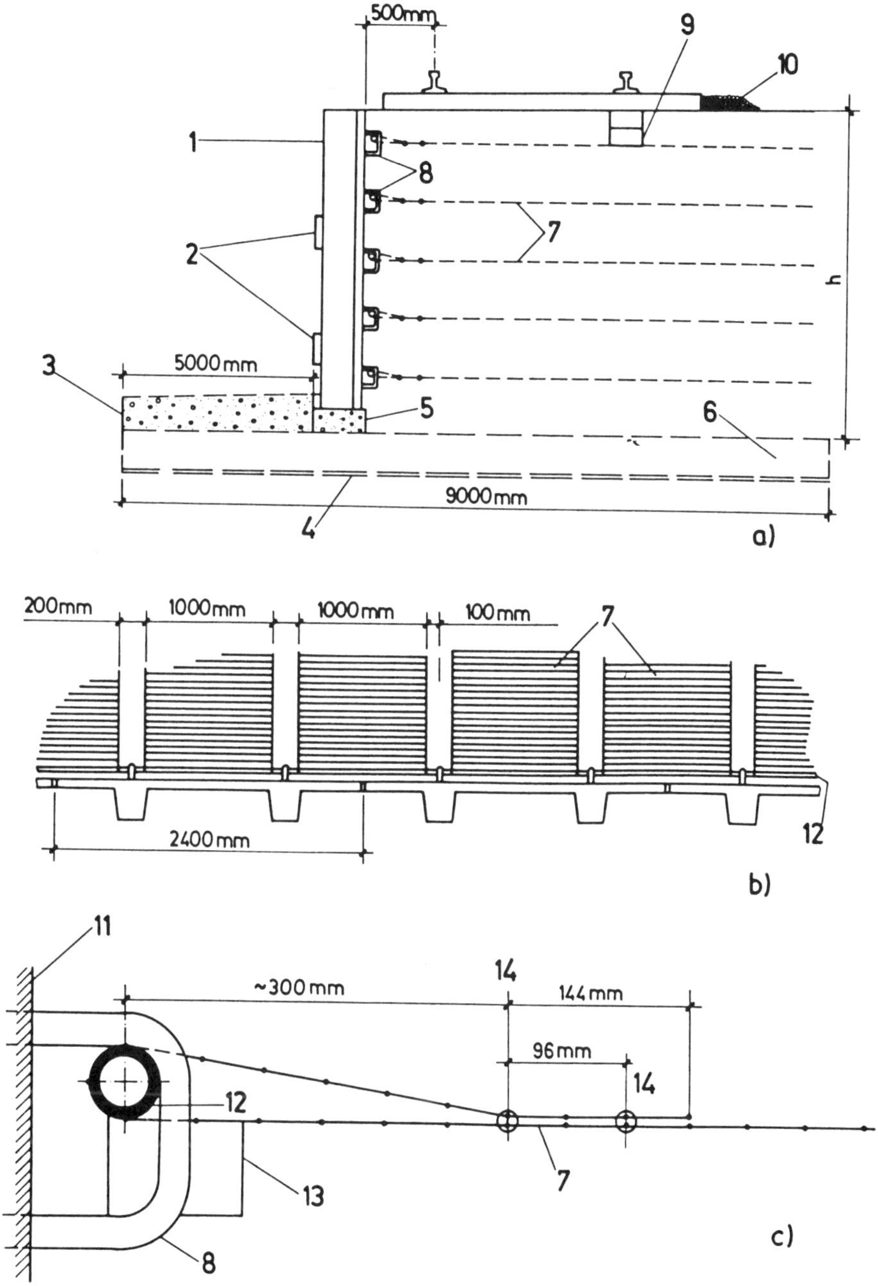

Figure 5–47 Plan and cross section of the reinforced soil structure at Newmarket Silkstone Colliery, together with details of the Tensar anchorage (*Rainbow, 1982*)
a) cross-section; *b)* plan; *c)* plan of the Tensar anchorage;
1 wall units; *2* timber fender units; *3* concrete slab; *4* layer of Netlon type CE 131; *5* concrete base; *6* limestone layer; *7* Tensar reinforcement; *8* steel bracket; *9* timber sleepers; *10* ballast; *11* precast wall unit with MS connectors; *12* 48.3 × 4 CHS; *13* foam rubber; *14* helix connector

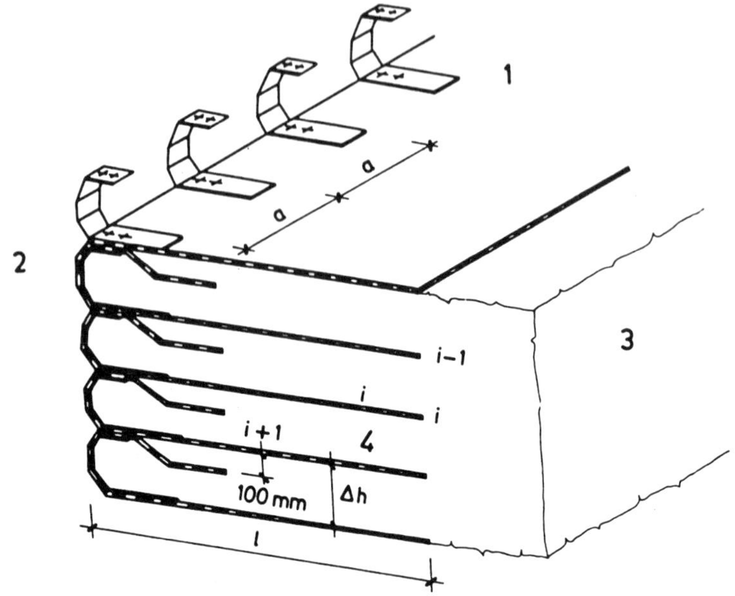

Figure 5–48 Elements of an earth retaining wall reinforced by geotextile (*Piesker, Göbel, 1982*)

1 metal stabilizer (frames for facing); *2* geotextile facing; *3* backfill; *4* reinforcing sheets

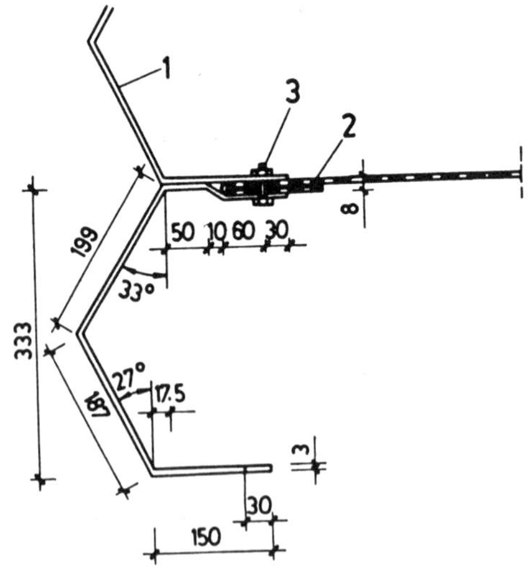

Figure 5–49 Connection of metal facing and geotextile

1 facing unit; *2* geotextile reinforcement; *3* bolts

A case in which the facing was formed by bending back the reinforcement sheets is shown in Fig. 5–48. Maintenance of shape is ensured by flat steel sheets folded in a regular polygonal shape (called local stabilization).

In Fig. 5–49, the connection between the structural elements of the wall and geotextile is shown (*Piesker, Göbel, 1982*). Connections can be made in other ways. One of them is worth mentioning here: in the 'shelf-wall', invented by *Broms* (*1977,* Fig. 7–14), an aesthetic appearance and effective connection is obtained by using rectangular elements placed on the edge of the fill. Connection between the reinforcement and the facing elements is ensured by friction between the horizontal plane and the reinforcement.

5.3.2 Stability analysis and design method for fabric reinforced earth walls

In the case of fabric reinforced earth walls, just as for conventional reinforced retaining walls the external and internal stability has to be checked. In this case, internal stability must be analysed using limit equilibrium methods. The problems of calculating for working conditions are as before. From the point of view of the backfill, these walls behave like gravity retaining walls, and, therefore, the external stability analyses are not repeated here.

In the internal stability analysis, the following dimensions must be calculated (Fig. 5–50):

— total length of the fabric (l),
— length of the fabric bent backwards (embedded length) (l_1),
— vertical spacing between fabric layers (f).

Design suggestions published earlier take the anchoring principle into account (*Bell, Steward, 1977*; *Murray, 1980*).

In analysis the ultimate tensile load per metre of fabric T_a is given. Corrosion need not be considered and because of the sheet-like geometry there is no horizontal spacing. These conditions simplify the situation in comparison with the reinforced earth structures. In the following, calculations developed based on *Lee*'s method and using analogous assumptions (including the simplifications presented earlier), are described (*Murray, 1980*).

A schematic illustration of a structure idealised as a gravity wall is shown in Fig. 5–51. The imaginary back of the wall is formed by the surface connecting the free ends of the sheets (vertically stepped or sloped). The active earth pressure resulting from the load of the backfill is

$$E_a - \frac{1}{2}\lambda_a h^2 \gamma \qquad (5\ 31)$$

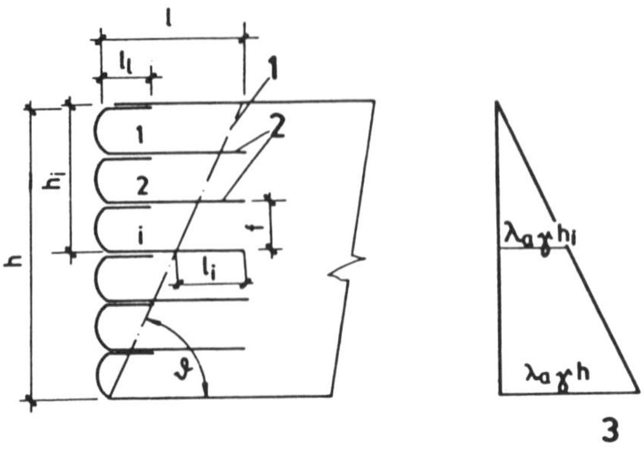

Figure 5–50 Interpretation of the geometrical data for internal stability analysis (*Murray, 1980*)

1 failure plane; *2* reinforcing fabric (sheet)

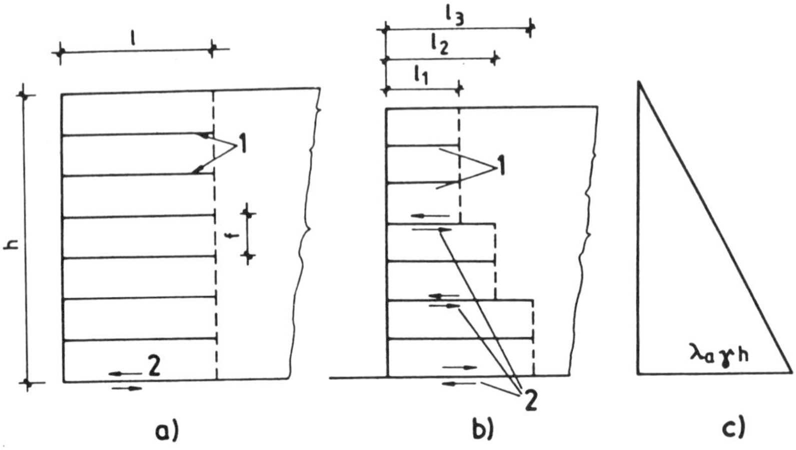

Figure 5–51 Different types of earth walls reinforced by sheets

a) reinforcement with constant length; *b)* reinforcement with variable length;

c) assumed pressure distribution;

1 reinforcement; *2* frictional resistance

where λ_a = coefficient of the active earth pressure, γ = unit weight of the soil and h = height of the wall. Assuming that the total length of the fabric l is equal at all levels, it can be assessed from the sliding criterion for the whole wall. The total weight of the wall is

$$G = hl\gamma \tag{5–32}$$

and if μ_s is the coefficient of friction between the soil and fabric, then the inequality

$$l \geq \frac{\lambda_a}{2} \cdot \frac{h}{\mu_s} \qquad (5\text{–}33)$$

is obtained. If the length of the fabric varies, then this calculation must be carried out at all levels.

For the total length l, another criterion must be specified: we define this from the criteria relating the equilibrium to the overturning of the wall and to earth failure beneath the front and the rear of the wall (Fig. 5–52).

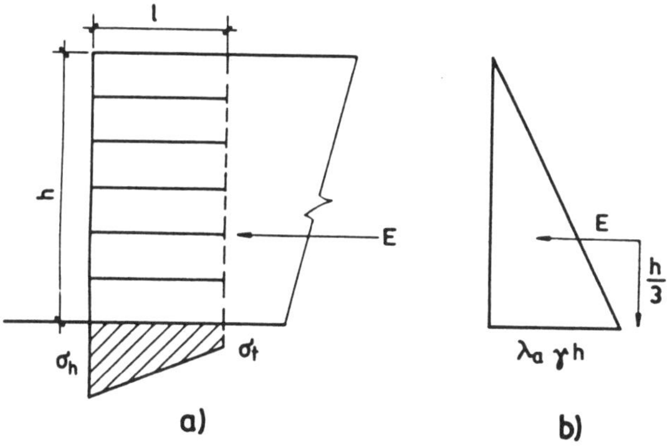

Figure 5–52 Vertical stress distribution beneath retaining walls (*Murray, 1980*)
a) vertical stress; *b)* horizontal earth pressure

For the stresses σ_h and σ_t beneath the front and the rear of the wall the inequalities

$$\sigma_h \leq \sigma_{tH} \qquad (5\text{–}34a)$$

and

$$\sigma_t \geq 0 \qquad (5\text{–}34b)$$

must be satisfied, in which σ_{tH} is the allowable soil bearing pressure. By assuming a linear stress distribution (a rigid body) the stresses are:

$$\sigma_h = \gamma h \left(1 + \lambda_a \frac{h^2}{l^2}\right) \qquad (5\text{–}35a)$$

and

$$\sigma_t = \gamma h \left(1 - \lambda_a \frac{h^2}{l^2}\right) . \qquad (5\text{–}35b)$$

From Eqs 5–35a and 5–35b two criteria can be expressed by these two inequalities:

$$l = \sqrt{\frac{\lambda_a \gamma h^3}{\sigma_{tH} - \gamma h}} \tag{5–36a}$$

and

$$l \geq h\sqrt{\lambda_a}. \tag{5–36b}$$

For the length l, the three criteria of Eqs (5–33) and (5–36) are derived from the analysis of external stability. The fourth criterion (generally it gives a larger value) is obtained from checking of internal stability. It is assumed that the failure plane is inclined at an angle $\vartheta = 45° + \varphi/2$ to the horizontal direction behind which the fabric transmits the force to the soil along the length l_i by means of adhesion. Let the force acting on the fabric at the level i at the depth h_i be

$$T_{fi} = 2\gamma h_i l_i \mu_s \tag{5–37}$$

where

$$l_i = \left[l - (h - h_i)\cot\vartheta\right] \tag{5–38a}$$

and

$$h_i = if. \tag{5–38b}$$

From the equilibrium of the horizontal stress components for the whole wall

$$\frac{1}{2}\lambda_a h^2 \gamma = \sum_{i=1}^{n} T_{fi}, \tag{5–39}$$

where T_{fi} is the force distributed on the surface of the fabric layers.

From the sum on the right hand side, the expression

$$\gamma \mu_s h^2 (n+1)\left[\frac{l}{h} - \frac{\cot\vartheta}{3}\left(1 - \frac{1}{n}\right)\right]$$

is obtained from which the inequality

$$l > h\frac{\lambda_a}{2\mu_s(n+1)} + \frac{\cot\vartheta}{3}\left(1 - \frac{1}{n}\right) \tag{5–40}$$

can be derived. Alternatively, a good estimate of length for large values of n can be obtained by rearranging Eq. (5–40) as follows:

$$l > \lambda_a h\frac{1}{2\mu_s n} + \frac{1}{3\sqrt{\lambda_a}}. \tag{5–41}$$

When determining the length, the safety factor is neglected because its value depends on many additional considerations: it can be defined as both the quotient of the forces in the equilibrium state and the coefficient of the dimensions arrived at by computation. The vertical spacing f between the fabric layers can be obtained from the condition of equilibrium relating the tension T_{fi} permitted in the reinforcement at the level i and the force acting on the fabric:

$$f_i = \frac{T_{fi}}{h_i \lambda_a \gamma} \tag{5–42}$$

again omitting the safety factor. It is generally accepted that the vertical spacing determined at the lowest level of the wall should be adopted everywhere (also assumed when calculating the limits of length) i.e.

$$f_n = \frac{T_{fi}}{\lambda_a \gamma h}. \tag{5–43}$$

In the internal stability analysis, the embedment length of the fabric l_1 must be checked. The critical or maximum tensile force acts on the lowest fabric:

$$\frac{1}{2}\lambda_a \gamma h f \tag{5–44a}$$

and the pull-out force is balanced by the frictional resistance:

$$\gamma(h - f)\mu_e l_1 \tag{5–44b}$$

where μ_e denotes the frictional coefficient along the interface of the fabrics. Since $h = nf$, the inequality

$$l_1 > \frac{\lambda_a h}{2(n-1)\mu_e} \tag{5–45}$$

is obtained which, in the case of a large value of n, can be approximated by

$$l_1 > \frac{\lambda_a f}{2\mu_e}. \tag{5–46}$$

In this step, the safety factor can be introduced as a coefficient in both Eqs $(5 - 44a)$ and $(5 - 46)$.

To determine the serviceability of the structure based on considerations of strain in the fabric, *Murray* suggests a simple method. He assumes that the stress in the reinforcing fabric is a maximum at the facing and decreases linearly towards the end of the reinforcement. Tension along the length in the lowest fabric can be expressed as follows:

$$T_{\mathrm{f}n} = \lambda_{\mathrm{a}} h \gamma f \left(1 - \frac{x}{l}\right), \tag{5–47}$$

where x is the distance from the facing.

If the reinforcement behaves as a linear elastic material, then its extension can be calculated from the following formula:

$$\Delta l = \int_0^l \varepsilon_{\mathrm{x}}\, dx = \int_0^l \frac{T_{\mathrm{f}n}}{E}\, dx \tag{5–48}$$

where E is the Young's modulus for the fabric, and for sheet-like fabrics it can be obtained from the tensile force–extension diagram for unit width. Using Eq. (5 – 48) we obtain the expression

$$\Delta l = \frac{1}{2E} \gamma h l f \lambda_{\mathrm{a}} \tag{5–49}$$

in which every parameter influencing the deformation of the facing is presented.

Naturally, Eq. (5 – 49) gives only an approximate value for deformation. This method can be improved by applying a more accurate stress–strain relationship (e.g. *Murray* suggests *Kondner*'s hyperbolic relationship; *Kondner, 1963*), and non-elastic effects, e.g. a hysteresis loop can also be considered using the method discussed in Chapter 3. It is not proposed to use this, or more sophisticated mechanical models, because some starting assumptions are uncertain (e.g. the distribution of the tensile stress). In this regard, we remind the reader of the remarks made earlier regarding numerical methods.

6

REVETMENT WALLS STRENGTHENED WITH DISTRIBUTED INCLUSIONS

Economy, lack of space, or technical problems (such as the hazard of crumbling, rolling, etc. of soils or steep inclinations of slopes) often require the construction of revetment walls in cuttings. Traditional retaining wall constructions can be applied in these cases, too, by temporarily excavating the full space necessary for the supporting structure. Novel technologies of trenching, anchoring or nailing, and advanced design methods can be used to produce purpose-made reinforced earth revetment walls as well. It is worthwhile recalling the advantages of this method in terms of safety and economy (only temporary supports are required, earth refill, compaction, can be avoided, etc.).

While we shall often refer to the importance of anchoring technology, this chapter does not deal with conventional anchored (rigid or flexible) retaining walls. Their mechanical behaviour and, particularly, their mathematical modelling (solid facing fixed into an undisturbed soil base by long, singular rods)—has very little to do with soil reinforcement. A widely distributed arrangement of inclusions performed by a similar technology results in 'nailed soil'. Another method of implanting thin prefabricated (or made in-situ) panels in the backing (in some sense similar to nailing) is closely related to trench wall technology. These reinforcing variants will be treated in detail in this chapter.

6.1 REVETMENT WALLS STRENGTHENED WITH SOIL NAILS

6.1.1 Structure and construction

The construction process for a vertical or slanting nailed soil revetment wall begins with excavation of the upper layer of the earth mass. The depth of this layer (Fig. 6–1*a*) depends on the free soil slope stability with respect to the time period required for construction. Then, the free surface is covered by some facing: the most commonly used method is shotcrete facing (Fig. 6–1*b*).

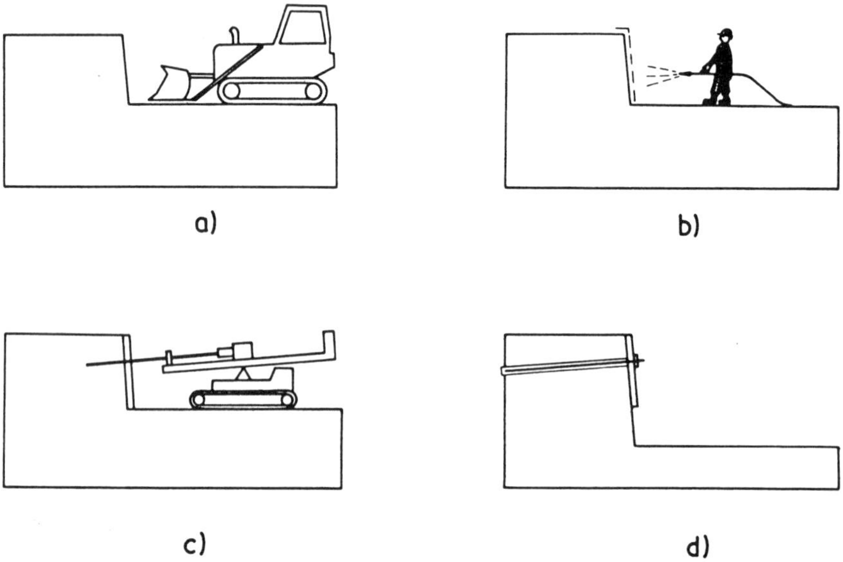

Figure 6–1 Construction scheme of revetment walls reinforced with soil nailing
a) earth excavation; *b)* making of shotcrete facing with steel mesh;
c) hole drilling, soil nail insertion; *d)* excavation of the next soil layer

The thickness of the facing may vary from 100...150 mm (temporary walls) up to 250 mm (permanent walls; *Stocker, Gässler, 1979*). Precast concrete panels can also be used for the same purpose. In this case, the space behind the panels is filled with coarse granular soil to serve as a filter, or a cement mortar injection is used to give a perfect seal.

Once the facing is in place, 'soil nails' are inserted into the soil (Fig. 6–1*c*). In current practice, steel or plastic rods of diameter 20...30 mm are embedded, usually by boring. Other techniques such as driving, pressing, vibrating are less frequently used. The crucial requirement

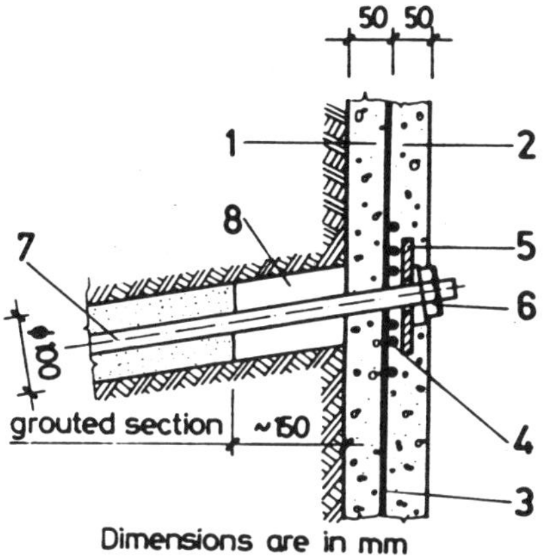

Figure 6–2 Detail of stressed soil nailing in the USA
1 first layer of shotcrete facing; *2* second layer of shotcrete facing; *3* steel reinforcement mesh, at 200 mm centres two-ways; *4* steel bars, horizontally laid; *5* steel plate, 150 x 150 x 12 mm; *6* M 24 washer and nut; *7* 24 mm ϕ steel bar; *8* ungrouted section

good adhesion between the soil and the nail along its full length. Injection of the borehole with cement mortar is the most effective way to develop perfect bonding between the soil mass and the reinforcing inclusions. The heads of the nails rest on the facing. A steel plate or rod mesh can improve the stress distribution (Fig. 6–2). The nails, which are fixed by nuts and washers, may be stress-free (European practice) or stressed up to 50...80% of the designed tensile force (US-practice,

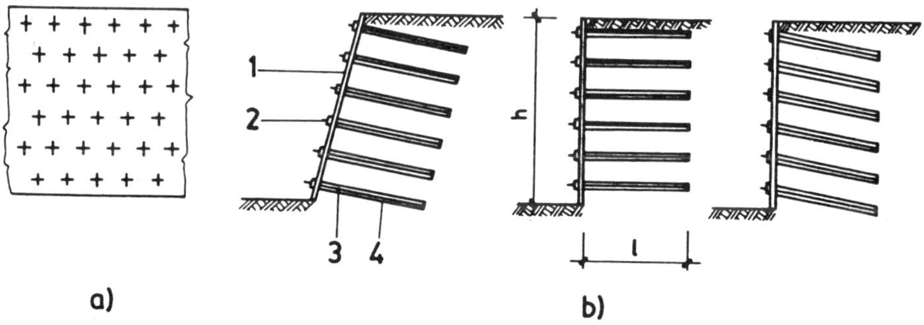

Figure 6–3 General arrangement of a revetment wall with soil nailing
a) elevation; *b)* cross section variants;
1 facing (shotcrete, precast r.c. elements etc.); *2* anchor head; *3* anchor rod soil nail (steel bar); *4* cement mortar grout

Bányai, 1982). If stressing is carried out, the last 150 mm of the nail must be free of mortar in order to avoid unwanted cracking in this region.

The technique is repetitive; the next step being the excavation of another earth layer (Fig. 6–1*d*). The length of the nails, depending on the soil parameters, varies between 0.5...0.6 h (h denotes the height of the wall) and the number of the nails is usually around 1–2 nails per m^2.

Several nailed earth wall arrangements are shown in Fig. 6–3. The wall shown in Fig. 6–4 was designed and built under difficult soil conditions in northern Hungary (*Gérusz, 1981*). The precast reinforced concrete plate facing elements of 1.5 m × 3.0 m are joined along notched joints. Each plate was fixed by two nails of length of 6...8 m. Behind the plates, coarse gravel (mixed with 30...50 kg/m^3 of cement) served as a drainage layer. The wall proved to be cheaper by 25...30% compared with conventional solutions.

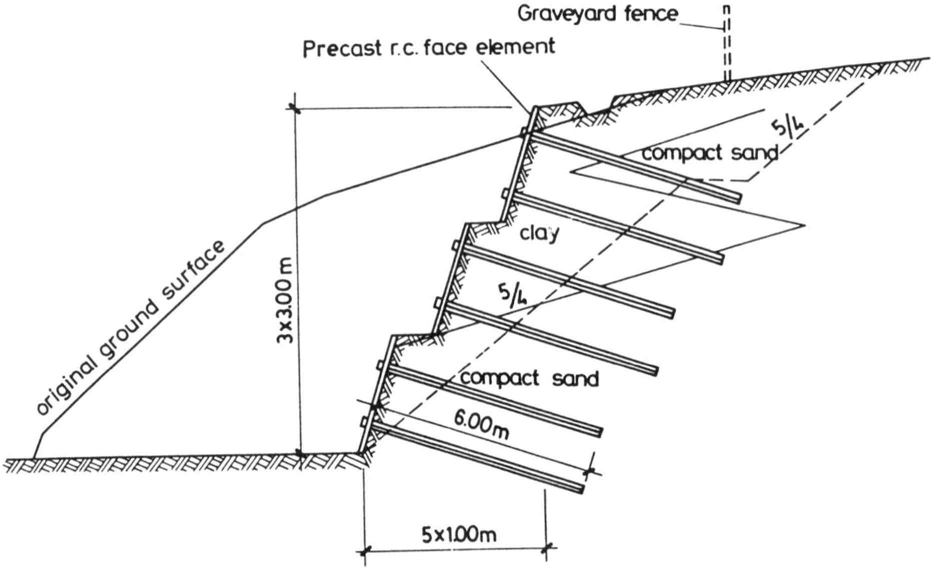

Figure 6–4 Revetment wall with soil nailing in Hungary (Design by FTV)

An example of a nailed soil revetment wall built near Paris is shown in Fig. 6–5 (*Cartier, Gigan, 1983*). The phases of the construction were as follows:

- earth excavation down to 1.4 m depth,
- placing the angled steel nails of dimensions (legs and thickness) 50–50–5 mm and 60–60–6 mm with lengths of 5.5 m and 7.0 m,
- facing with a 0.05...0.10 m thick shotcrete layer reinforced with welded steel net (the nail heads were covered by the concrete facing).

The soil mass consisted of fine sand described by the following parameters:

water content 14 ... 22%,
specific weight 20 ... 21 kN/m^3,
particles smaller than 80μm less than 0.5%,
angle of internal friction,
(measured in shear box) 30 ... 36°,
limit pressure from pressuremeter 1.8 MPa,
pressuremeter modulus 16 MPa,
(these data reflect a relatively dense state).

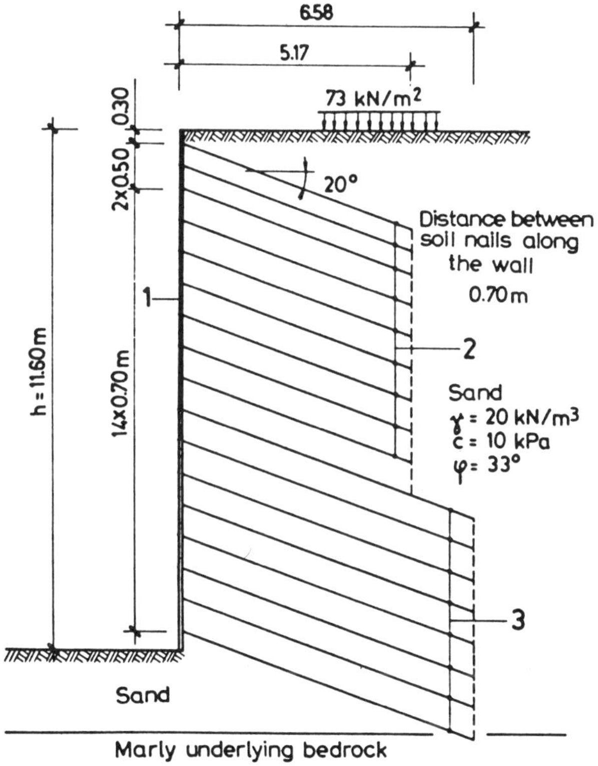

Figure 6–5 Revetment wall with soil nailing, near to Paris (Hurpinoise-wall; after *Cartier and Gigan*)

1 facing; shotcrete 0.05 ... 0.10 m thickness, with welded steel reinforcing mesh;

2 9 soil nails made of angled steel 50 × 50 × 5 mm, of 5.5 m length;

3 8 soil nails made of angled steel 60 × 60 × 6 mm, of 7.0 m length

During the construction period (less than four months), 14 mm of vertical settlement was observed at the upper edge of the wall. This value did not increase subsequently. Experimental data for the nail-soil adhesion force of 15 ... 20 kN/m were obtained by pull-out tests at different depths during the construction.

Further examples were published with valuable observations and experimental data in the Federal Republic of Germany, where more than 20 nailed soil walls have been erected until 1980.

6.1.2 Internal forces in a nailed soil wall

The nails—like distributed inclusions—carry the major proportion of the shear stresses in the soil mass by adhesion and friction on the soil, injected mortar and nail interfaces. In this way, the nails carry a tensile load while the pressure acting on the facing decreases. Therefore, nailing results in a reinforced soil mass which has greater shear resistance than the original soil and acts like a gravity wall and the nails crossing any potential slip surface have significant shear resistance and increase the stability of the wall.

6.1.2.1 Large-scale experiment in the Federal Republic of Germany

These general statements describing the mechanical behaviour of nailed soil walls were verified by extensive investigations performed by the Institute of Soil and Rock Mechanics, University of Karlsruhe, in cooperation with K. Bauer Ltd. A 7.0 m long experimental wall with the cross-section shown in Fig. 6–6 was built to retain (in sections) loose sand, medium dense sand, silt and clay (*Stocker, Gässler, 1979*; *Fejér, 1980*; *Bányai 1982*). The soils had the following parameters:

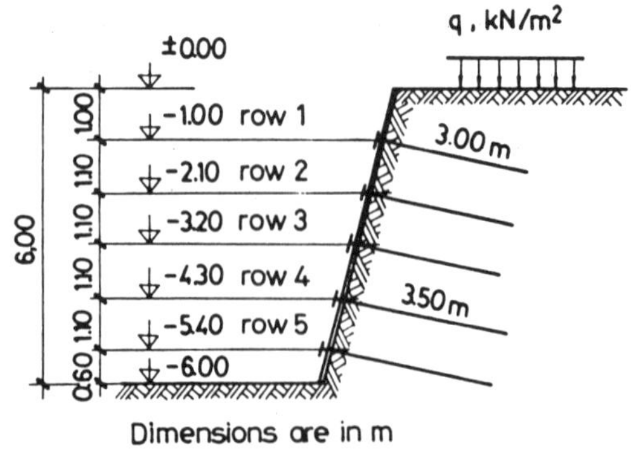

Figure 6–6 Cross section of an experimental retaining wall (*Stocker, Gässler, 1979*)

wet unit weight of the sand 15 ... 16 kN/m^3,
angle of internal friction (measured by shear tests) 42°,
undrained strength of the silty clay:

at the top	50 kN/m^2,
in the middle	160 kN/m^2,
at the bottom	280 kN/m^2 of the wall.

Several load intensities and load positions were investigated, including also dynamic loading. The following data were measured:

— horizontal displacement of the facing,
— settlement of the ground surface,
— displacements inside the soil wall,

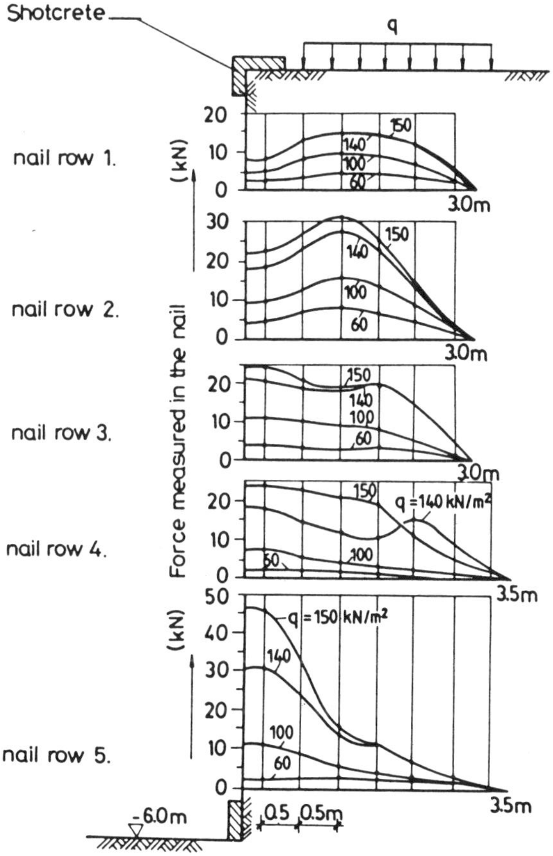

Figure 6–7 Distribution of nail forces measured in an experimental retaining wall (*Stocker, Gässler, 1979*)

— nail forces along the nails and at the facing,
— earth pressure acting at the shotcrete facing,
— vibrations in the soil and facing caused by the dynamic loads.

Results of the displacement measurements

During the 'from the top down to the bottom' building process the horizontal displacements reached a value of 0.15...0.22% of the height. Under relatively high additional surface loads the deformations remained in this range: in granular soil under a load of 100 kN/m^2, 0.1 percent was observed; in the wall section retaining the silty clay this value was 0.2 percent at the end of a consolidation period (about 220 hours).

To establish realistic failure mechanisms, the positions of the slip surfaces were also determined by using bored inclinometer measurements. Two characteristic patterns of the slip surfaces were found:

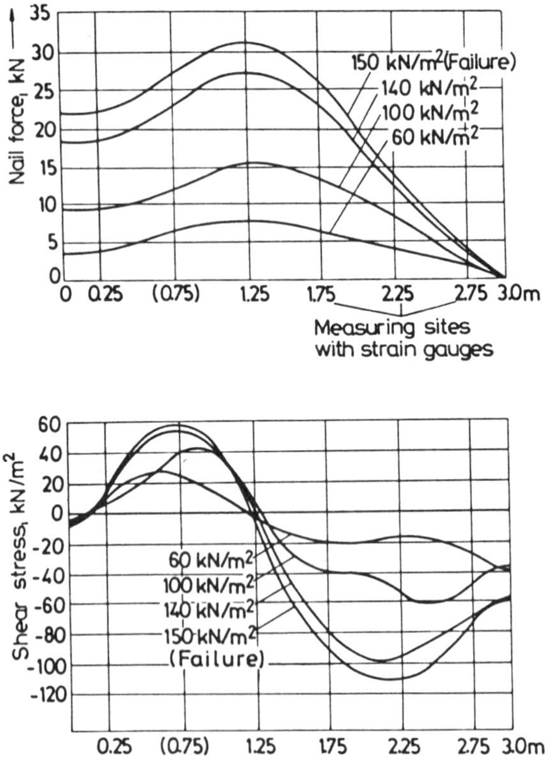

Figure 6–8 Tensile force distribution as measured in the 2nd nail row of the experimental retaining wall and distribution of shear stresses along the nail (*Stocker, Gässler, 1979*)

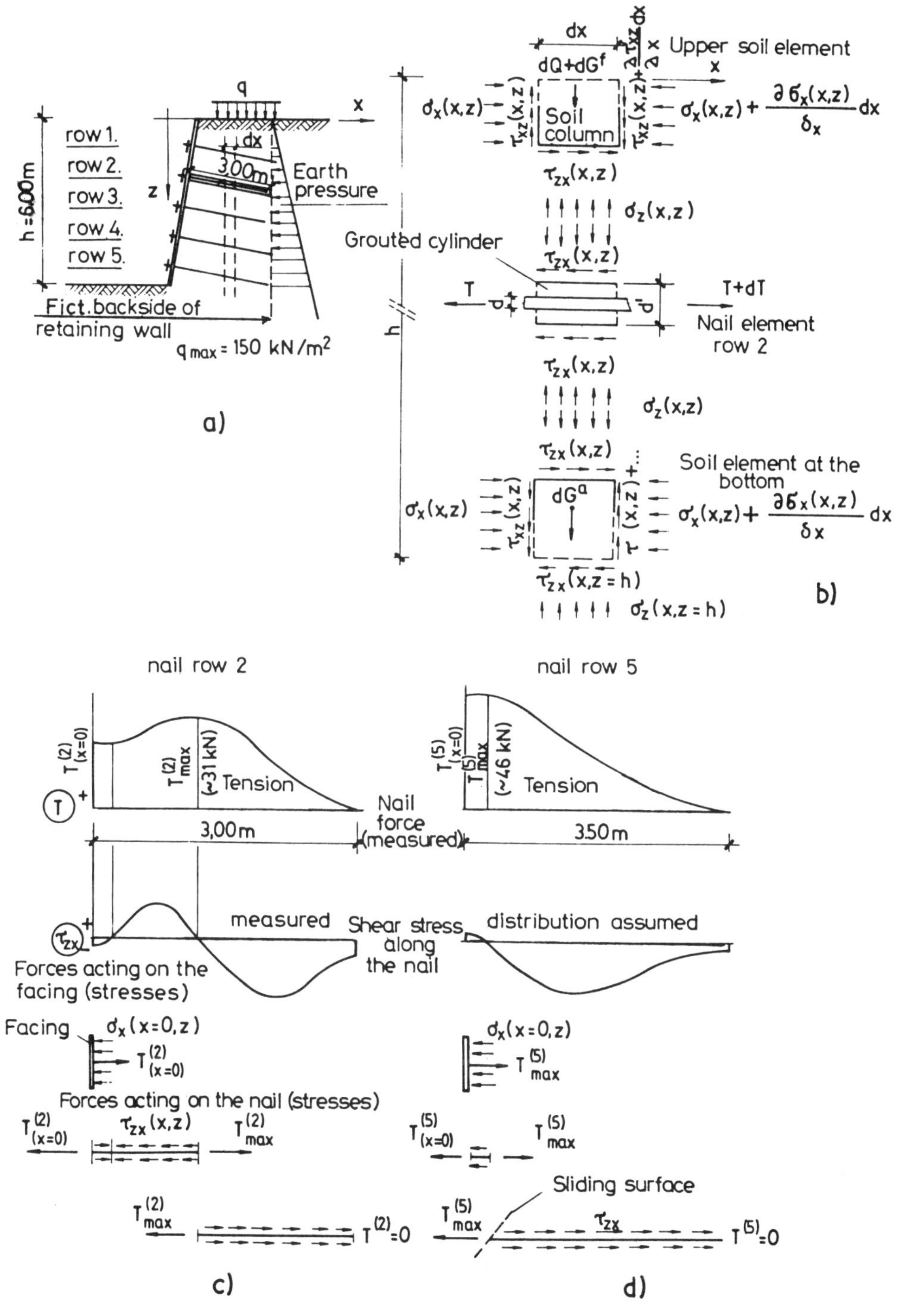

Figure 6–9 Internal forces in an experimental retaining wall with soil nailing

a) cross-section; *b)* equilibrium of a soil column of width dx;

c)... d) forces (stresses) acting on the nail rows

- under distributed loads starting at the wall edge, simple linear slip surfaces developed;
- under localized loads some distance from the edge of the wall, complex slip surfaces (compound failure mechanism) developed.

Earth pressure and nail force measurements

The earth pressure increased proportionally to depth, reaching a maximum at about the lower third of the wall height; below this level its value decreased.

The earth pressure resultant was not higher than 70% of the active *Rankine* force, even under the higher surface loads. The measured intensity of the earth pressure distribution turned out to be about half of the value obtained by conventional methods. The earth pressure at the facing, caused by the soil weight, was similarly lower.

The nail forces distribution measured along the nails in rows 1...5, are shown in Fig. 6–7. In the two upper rows not crossed by the failure surface, the maximum tensile force occurred in the middle section of the nails. Details of the measured data in the 2nd row can be seen in Fig. 6–8. They illustrate the unloading effect at the facing. The shear stresses acting at the nail–soil interface are higher than 100 kPa. In the lower rows the tensile force reaches its maximum at failure, at the point where the nail crosses the slip surface, with shear stresses as high as 200 kPa.

The equilibrium state of a soil column of width dx is shown in Fig. 6–9 (for the sake of clarity with one nail section only). Inserts *b* to *d* in the figure show details of the tensile force and shear stress distributions both at the facing and at the nail.

6.1.2.2 Large-scale experiments performed in France

An experimental wall with a height of 5.50 m was investigated in France (*Cartier, Gigan, 1983*). The tensile force distribution was measured along the angled steel nails (Fig. 6–10). Pull-out experiments were performed at the same wall with nails made of steel pipes as well, to determine the distribution of the tensile forces (Fig. 6–11) and the friction along the mortar–soil interface (Fig. 6–12).

The frictional resistance mobilized at the relative displacement level of 2...3 mm between the nail and the soil was about 100...120 kPa.

Cartier and *Gigan* found that the expression suggested by *Schlosser* and *Guilloux* for the mobilized specific frictional force acting on the interface, namely:

$$T_f = Kc + 2s\sigma_v\mu^* \qquad (6\text{–}1)$$

is a good approximation. Here μ^* denotes the apparent frictional coefficient, and the following notation is used:

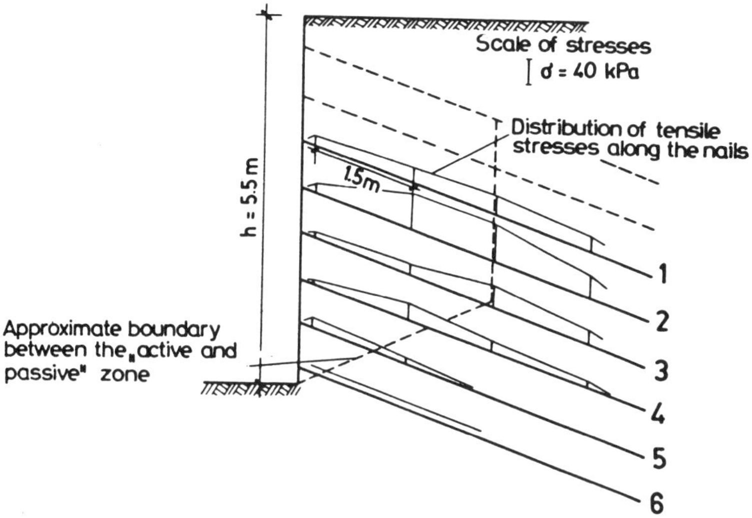

Figure 6–10 Distribution of tensile stresses in the soil nails. Experiment in France (after *Cartier and Gigan*)

Steel nails: 50 × 50 × 5 mm angled steel; length: 5.50 m; at 0.70 m centres

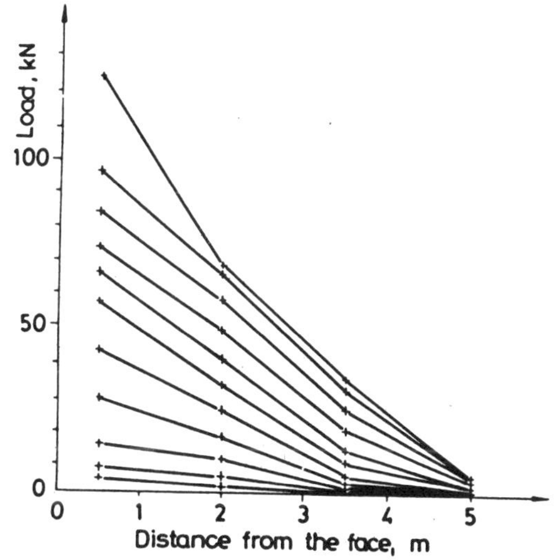

Figure 6–11 Tensile forces in a soil nail made of steel pipe, in a pull-out test for monotonically increased forces. Experiment in France (after *Cartier* and *Gigan*)

K — perimeter of the nail cross section,
c — cohesion of the soil,

s — lateral diameter of the reinforcing inclusion,
σ_v — vertical stress ($z\gamma$) at the average depth of the nail.

Values of $\mu^* > 1.0$ at the upper part of the wall reflected the dilatancy effect. At greater depths, for smooth inclusions $\mu^* = 0.5 \tan\phi$, and for inclusions with rough surfaces $\mu^* = \tan\phi$.

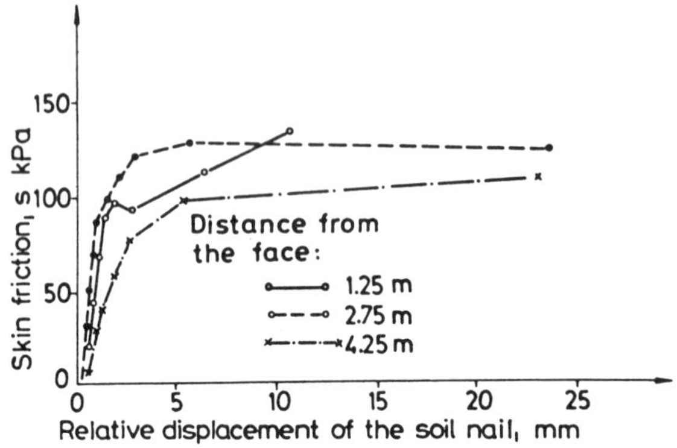

Figure 6–12 Skin friction along a soil nail made from steel pipe. Experiment in France (after *Cartier* and *Gigan*)

6.1.3 Design of nailed walls

6.1.3.1 Failure mechanism approach

In the Federal Republic of Germany, small scale laboratory experiments were performed to study the failure mechanisms. From the results published between 1979 and 1981, the most important design criteria can be divided into two groups:

- internal stability (internal failure mechanisms, nail forces, design of facing and nails);
- external stability (including translational, rotational, soil- and ground failure mechanisms) where the nailed soil mass is idealised as a single rigid block.

Investigation of the internal stability

Internal stability can be investigated by assuming internal failure surfaces. If this kind of stability is assured, checking the rotational mechanisms is not necessary because the nailed soil mass is safely fixed by the nails into the region behind the failure surface.

The safety of the wall system can be characterized by the minimum value of the safety coefficients obtained for those kinematic failure mechanisms observed in experiments (the safety factor determined from a kinematic approach is equal to or greater than the effective one and is therefore an upper bound). Small-scale experiments performed to define the internal failure mechanisms (*Gässler, Gudehus, 1983*) showed that (Fig. 6–13):

— at a freely rotating wall of h = 0.35 m height subjected to surface load, failure occurred at $q = 0.75\gamma h$ on a circular surface (Fig. 6–13*a*); because of the small curvature of this surface, it can be approximated by a plane,

— at a wall of height h = 0.45 m, fixed against rotation, and loaded by a horizontal force ΔC (Fig. 6–13*b*), the failure surface extends beyond the reinforced soil mass and can be approximated by two planes,

— at a freely rotating wall, h = 0.35 m, with partial surface loading acting outside the reinforced region, the failure occurred at $q_2 = 2.5\gamma h$. Three compatible curved surfaces were observed with the radii shown in Fig. 6–13*c*.

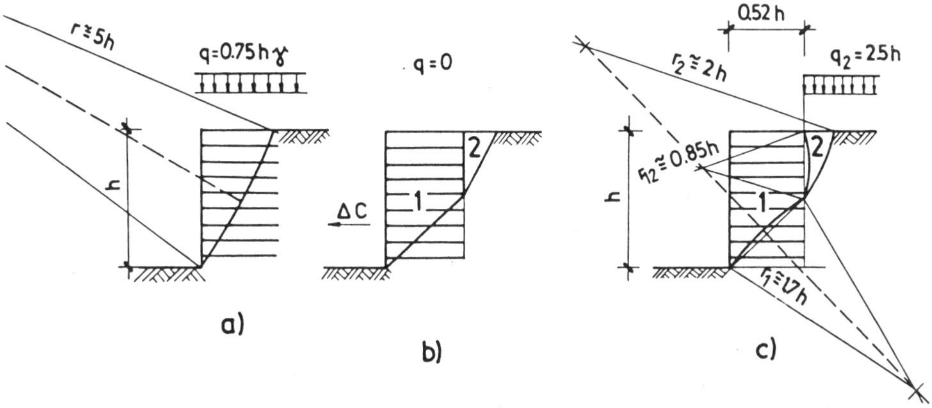

Figure 6–13 Failure mechanisms produced in small-scale experiments (*Gässler, Gudehus, 1983*)

a) simple, sliding circle; *b)* compound sliding planes; *c)* compound sliding circles

The highest safety factor was found in the case shown in Fig. 6–13*b*, but the differences were practically negligible for the range of loads $q < h\gamma$. It was concluded that the safety factor against failure can be calculated by considering only the two-plane compound mechanism. This analysis is described in what follows.

Investigation of the compound failure mechanism

(a) Determination of the safety factor from equilibrium conditions in cohesionless soil. In the case of loads distributed beyond the edge of the wall, the failure mechanism consists of two blocks as shown in Fig. 6–14.

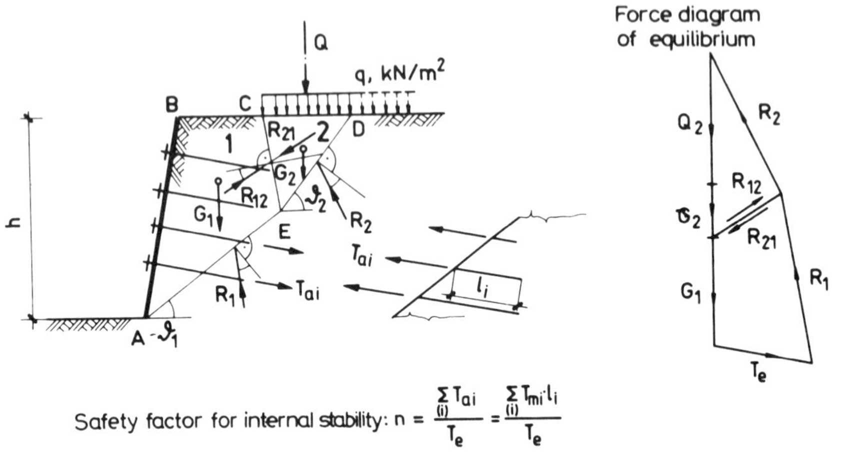

Figure 6–14 Compound failure mechanism and ultimate state of equilibrium (*Stocker, Gässler, 1979*)

This failure mechanism assumes that displacements are associated with two slip surfaces:

— the slip surface at an angle ϑ_1 develops in the nailed quadrilateral mass ABCE (1), which is subjected to no surface load,

— the slip surface inclined at an angle ϑ_2 forms the triangle CDE (2), where the load q acts on the ground surface.

The forces acting on the blocks in the limit equilibrium state are given in the force diagram, where $R_{12} = -R_{21}$ stands for the internal reactions acting along the secondary slip surface CE.

T_e denotes the resultant of the forces acting in the nails crossed by the slip surface AE. Deadweight and active forces are denoted by G_1, G_2 and Q_2, respectively.

The system satisfies the condition of internal stability if the resultant of the nail forces T_{ai} crossed by the slip surface AE exceeds the force T_e required to maintain equilibrium. Thus the factor of safety can be defined as:

$$n = \frac{\sum\limits_{(i)} T_{ai}}{T_e}. \tag{6–2}$$

The upper bound of the force T_{ai} is proportional to the specific shear force T_{mi} (kN/m) and nail length l_i (m):

$$T_{ai} = T_{mi} l_i \tag{6-3}$$

where T_{mi} depends on the soil–nail surface friction and should be determined on-site from pull-out experiments.

Assuming different failure mechanisms with different ϑ_1 and ϑ_2 angles, the minimum value of the factor of safety n can be found. The safety condition is satisfied if $1.5 < n_{min} < 2.0$. It was found from the experiments that dynamic loads (e.g. traffic loads) do not require more rigorous analysis.

(b) Soils with cohesion; kinematic approach. The kinematic approach (as described by *Gudehus, 1972*; *Goldscheider, Gudehus, 1974*) can also be used for nailed walls. Here, the example given by *Gässler* and *Gudehus* (*1981*) will be presented (Fig. 6–15).

Following experimental observations, two rigid soil blocks are considered. The slip surfaces are described by the angles ϑ_1, ϑ_2 and ϑ_{12} (Fig. 6–15*a*). These values can be selected independently, although they can also be interrelated using simple assumptions. Using a hodograph of arbitrary scale (Fig. 6–15*b*) the velocities v_1 and v_2 of the relative motions of the two soil blocks can be obtained (in the example $v_{1x} = 1$ was taken).

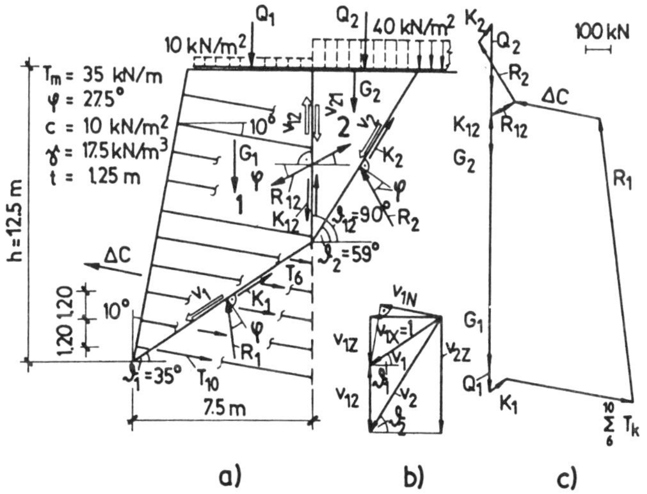

Figure 6–15 Determination of safety against failure in a nailed wall using the kinematic approach (*Gässler* and *Gudehus* 1981)
a) failure mechanism; *b)* hodograph; *c)* vector diagram

To determine the polygon (shown in Fig. 6–15*c*) of the equilibrium forces the resultant of the nail forces $\sum_{k=6}^{k=10} T_k$ can be calculated by the method described above. The cohesion forces K_1, K_2, K_{12} are determined by the cohesion (or apparent cohesion) value c, and φ denotes the angle of ultimate internal friction.

The equilibrium condition is related to the fictitious force ΔC. The simplest assumption of a force acting parallel to the nails is shown in Fig. 6–15*c*. The overall safety against failure can be defined in this case as:

$$n = \frac{W}{W_k} \qquad (6\text{–}4)$$

where W denotes the work done (or dissipated) by the forces acting along the slip surfaces and in the nails, and W_k stands for the work done by the external forces. This is the same concept of safety as that often used in slope and shallow foundation stability problems. Using the notation introduced in Fig. 6–15:

$$W = -\sum_{i=6}^{i=10} T_i v_{1N} - (K_1 + R_1 \sin\varphi)v_1 - (K_2 + R_2 \sin\varphi)v_2, \qquad (6\text{–}5)$$

$$W_k = (G_1 + Q_1)v_{12} + (G_2 + Q_2)v_{22}. \qquad (6\text{–}6)$$

(The work performed by the fictitious force ΔC is of course omitted here; taking this into consideration would result in $W = W_k$).

In the given example, the value $n = 1.20$ was obtained for the safety factor. The result, certainly, depends on the definition of n and ΔC. Partial safety factors and statistical methods could also be introduced. The calculation is repeated with different failure mechanisms to find the minimum value of n. Several numerical investigations have shown that for similar failure mechanisms, the minimum value of the safety factor can be well approximated by fixing

$$\vartheta_2 = 45° + \frac{\varphi}{2} \qquad \text{and} \qquad \vartheta_{12} = 90°,$$

so that only the effect of ϑ_1 need be considered.

Of course, further mechanisms of failure involving cylindrical, multiplanar or combined surfaces, as shown in Fig. 6–16 could be considered. Several investigations performed by using these assumptions have proved that in all cases the minimum value of safety factor determined for the 'double plane' failure mechanisms (Fig. 6–15), which can be easily calculated, serves as a good approximation to the actual minimum. Since

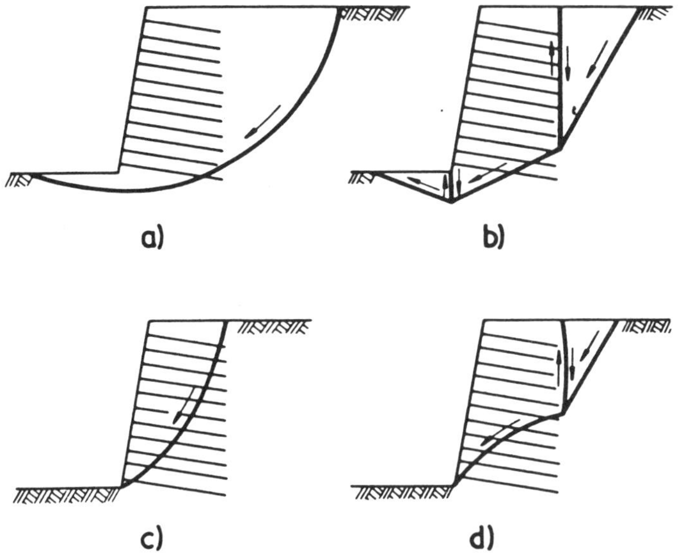

Figure 6–16 Analysis of various failure mechanisms (*Gässler, Gudehus 1981*)
a) deep sliding circle; *b)* deep compound translational failure mechanism; *c)* steep sliding circle; *d)* compound tilting mechanism

this assumption is good enough for design purposes, no further stability analysis are required.

(c) Practical design method. If we accept the results described above, the calculations can be easily codified and computer programs can be developed. Charts have been published which show the influence of the following variables (Fig. 6–17):

— inclination angle of the wall: $\alpha = 5° \ldots 20°$,
— slope angle of the ground surface: $\varepsilon \leq 15°$,
— length of the nails: $l = (0.6 \ldots 0.7)h$,
— surface loads: q_1, q_2,
— horizontal distance between the nails (t) and in vertical direction (f),
— inclination angle of the nails: β,
— inclination angle of the nailed soil block at its back: ρ,
— parameters of the soil (angle of internal friction φ, cohesion c, specific bulk weight γ),
— average shear force T_m mobilized along unit nail length.

Minimum values of the factor of safety n computed as a function of dimensionless variables are shown in Fig. 6–18. Here, the cases of $l/h = 0.6$ and 0.7 are plotted with the parameters $c = 0$, $\alpha = 10°$, $\varepsilon = 0°$,

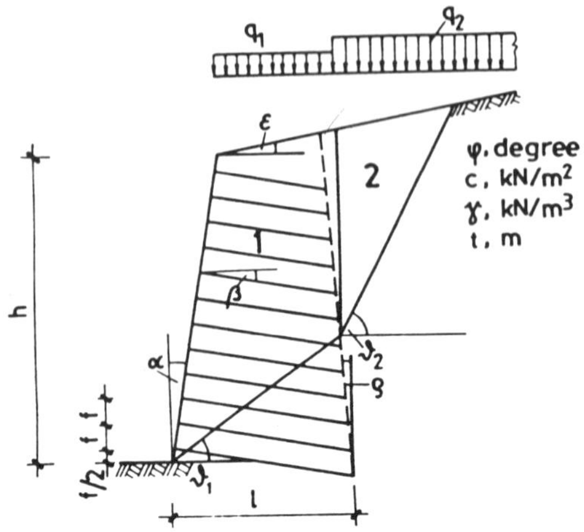

Figure 6–17 Scheme and notation of a nailed wall (*Gässler, Gudehus, 1981*)

$\beta = 10°$, $\rho = 0°$, $q_1 = q_2 = 0$. The required specific nail force $T_f = T/\gamma t f$ can be determined as a function of φ and prescribed n.

The suggested design method is as follows. The parameters of the soil mass (γ, φ, c) are to be determined by the usual methods as for other conventional retaining walls (weighted averages can be used for stratified soil masses). The values of φ and c should be obtained from laboratory and on-site tests or observations. Pull-out tests are required to determine the specific nail force T_m. Having prescribed the global safety factor n, the safe spacing data for the nails can be obtained from design charts (such as Fig. 6–18). The cross section can be checked using the ϑ_1 values given in the diagrams, if required. For example, if $n_{min} = 1.25$ and $\varphi = 30°$. The characteristic values relating to these data, obtained from Fig. 6–18*a* are $\vartheta_1 = 30°$ and $T_f = 1.36$. Given the value of T_m, the spacings f and t can be calculated.

The economy of the arrangement depends primarily on T_m and n. With greater experience, higher values of T_m and smaller values of n can be used. Some additional increase in safety can be achieved with reduced values of φ and c.

Simplified failure mechanism

In the experiments performed in the Federal Republic of Germany, approximately linear slip surfaces were observed under uniform loads from the upper wall edge (including the trivial case of the unloaded surface). In these cases safety factors can be determined in the same way as for

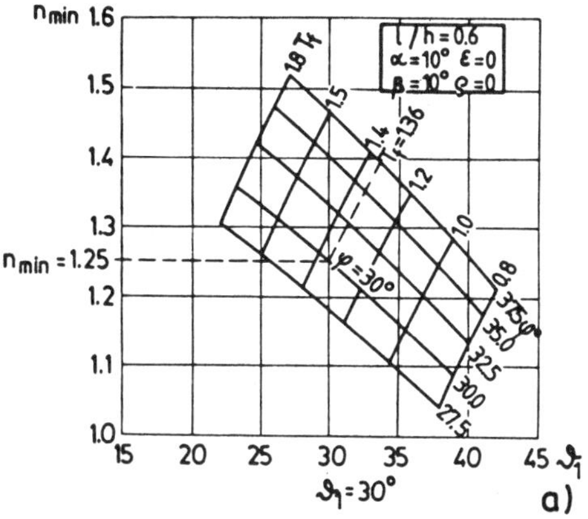

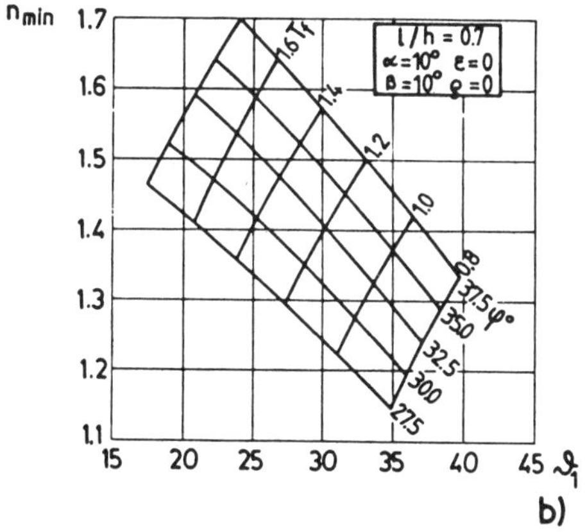

Figure 6–18 Diagrams for the design of a nailed wall complying with the general scheme in Fig. 6–17 (*Gässler, Gudehus, 1981*)

Relationship between the minimal factor of safety, inclination of the sliding surface, angle of friction and specific nail force T_f;

n_{min} minimal factor of safety;

ϑ_1 inclination of the sliding surface corresponding to the minimal factor of safety

the compound mechanisms with minor simplifications and changes of variables, as shown in Fig. 6–19.

Investigation of external stability

As noted earlier, the investigation of external stability can be omitted if the minimum value of safety has been determined using compound failure mechanisms. Nevertheless, it may happen (for example, if the nailed wall rests on a soft layer) that design codes or directives require proof of external stability.

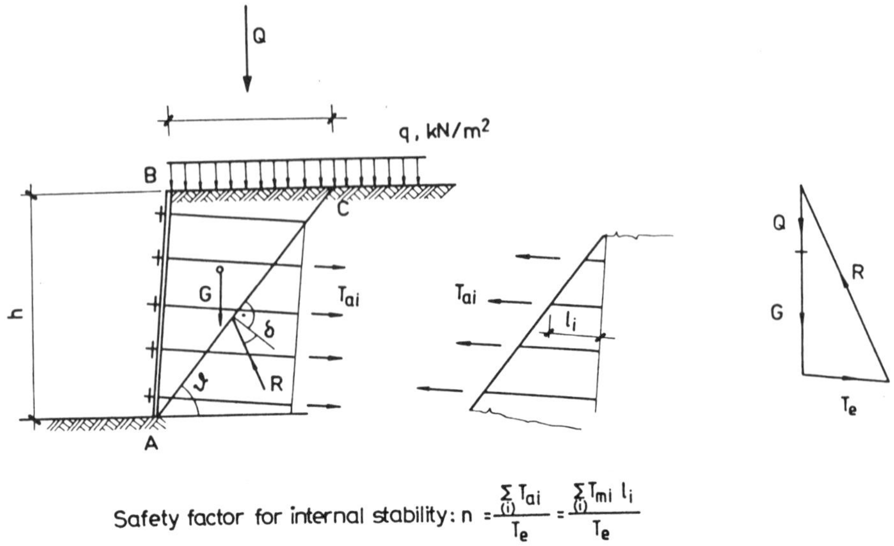

Figure 6–19 Simple failure mechanism and ultimate state of equilibrium

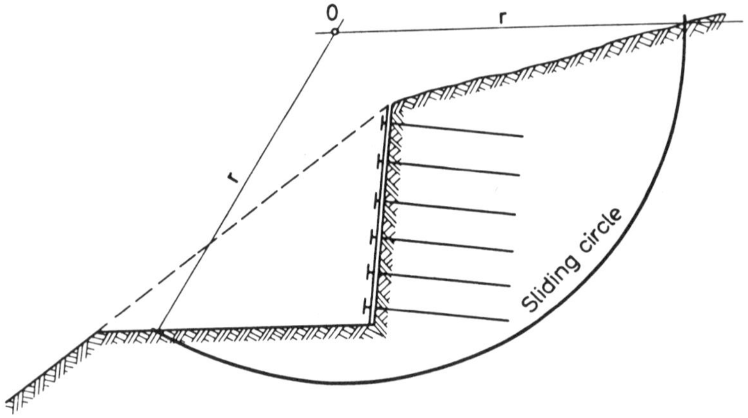

Figure 6–20 Assumed sliding circle for the determination of safety against ground failure (Investigation of external stability)

In these cases, the traditional methods can be applied again, i.e. idealizing the nailed wall as a monolithic mass. German standards suggest, for example, a factor of safety $n > 1.5$ (1.3) against slip (DIN 1054). The data in brackets refer to temporary walls or to walls under construction. The recommended value of factor of safety against foundation failure (as shown in Fig. 4–2*b*) given in DIN 1054 and DIN 4017 is $n > 2.0$ (1.5).

In the case of ground failure (Fig. 4–2*d*, Fig. 6–20) DIN 4084 suggests $n > 1.4$ (1.3). As a rule of thumb in geotechnics, tensile soil stresses are excluded; therefore the condition $e < b/6$ ($b/3$) is suggested for resultant force eccentricity by DIN 1054 (b denotes the wall width).

6.1.3.2 Method of design suggested by the Hungarian Company for Geodesy and Soil Investigation (FTV)

The nailed soil wall can be designed by the same method used for reinforced earth walls. This idea was adopted in the example mentioned before and shown in Fig. 6–4 (*Gérusz, 1981*). The general scheme of this method can be seen in Fig. 6–21, applied to soils with cohesion.

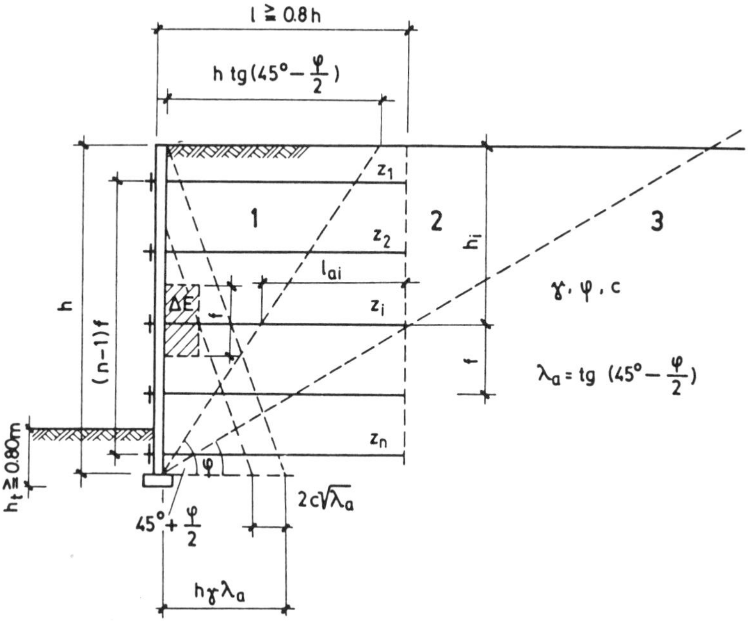

Figure 6–21 Design model for a nailed soil retaining wall as suggested by FTV
1 sliding soil zone; *2* intermediate soil zone, where certain displacements may occur; *3* stable soil zone

The earth pressure acting on the facing in this case can be approximated by the active *Rankine* distribution making allowance for the presence of cohesion. The nail forces T_i can be obtained from the average lateral earth pressure resultants E_i for each nail row:

$$T_i = \Delta E_i = \left(h_i \gamma \lambda_a - 2c\sqrt{\lambda_a}\right) ft, \qquad (6\text{–}7)$$

where

$$\lambda_a = \tan^2\left(45^\circ - \frac{\varphi}{2}\right),$$

and the following notation is used :

h_i — depth of the i-th row of nails,
γ — specific weight of the soil,
c — cohesion,
f — vertical distance between the rows of nails,
t — lateral distance between the columns of nails.

The force resisting pull-out of the nails can be obtained from the frictional forces at the nail–soil interface:

$$T^{\mathrm{s}}_{\mathrm{a}i} = d\pi l_{\mathrm{a}i} h_i \gamma \tan\varphi, \tag{6–8}$$

and the adhesion force:

$$T^{\mathrm{c}}_{\mathrm{a}i} = d\pi l_{\mathrm{a}i} c, \tag{6–9}$$

(here d represents the diameter of the nail, $l_{\mathrm{a}i}$ is the nail length; Fig. 6–21). To increase the safety, instead of soil cohesion adhesion between the nail and the soil

$$c_{\mathrm{a}} = \frac{2}{3}c \tag{6–10}$$

can be taken. Therefore, the pull-out resistance of the nail in the i-th row becomes:

$$T_{\mathrm{a}i} = T^{\mathrm{s}}_{\mathrm{a}i} + T^{\mathrm{c}}_{\mathrm{a}i} = d\pi l_{\mathrm{a}i}(h_i \gamma \tan\varphi + c_{\mathrm{a}}) \tag{6–11}$$

and $T_{\mathrm{a}i} > T_i$ is a plausible safety requirement.

6.1.3.3 Design of the facing and the nails

The load acting on the facing—in accordance wit the experimental results presented in Section 6.1.2—can be taken as half of the active *Rankine* earth pressure value.

The shotcrete facing can be modelled as a continuous plate supported at distinct points. Thus, the facing must also be checked for point load resistance.

The soil nails are designed for the maximum tensile forces obtained from the internal stability condition. Of course, additional design aspects (connected, for example, with corrosion) should also be considered. These questions are common in engineering practice and need no further comment.

6.2 SOIL WALLS REINFORCED WITH THIN PANELS

Strip reinforced soil walls (discussed in detail in Chapter 5) are probably the most economic configuration since both inclusions and soil are stressed. The situation is less obvious when the reinforced earth technique does not conform to the final wall geometry. In the case of walls built into cuts, a certain redundancy arises in removing and backfilling the soil; the extra costs may be significant.

6.2.1 Construction and building technology

The structures described in this section embody soil–structure interaction principles which are related to the well-known silo effect. To illustrate these principles, we consider the geometry of the problem. Narrow walls (with a thickness of 0.1…0.2 m) are inserted into the soil before any other site work is performed. These panels interact with the soil surrounding them and form a soil–structure system working as a composite gravity wall. The interaction depends on the dimensions and soil characteristics.

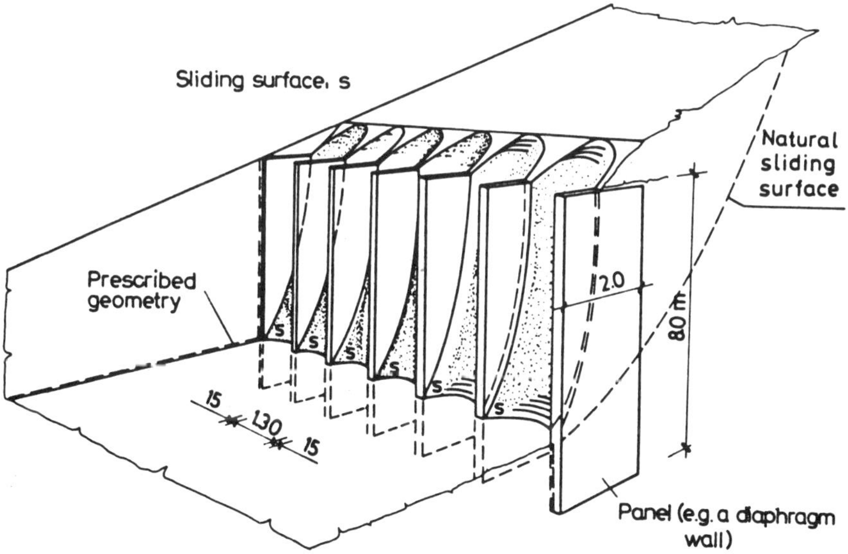

Figure 6–22 Axonometric view of a revetment wall strengthened with panels (without facing, or with facing of the sliding surfaces between the panels, say by shotcrete)

Having placed the panels, the soil is removed from the space in front of the wall. Assuming a granular soil mass, and no rigid, fixed facing between the panels, double curved surfaces develop as a result of the internal soil friction and the soil–wall surface interaction (Fig. 6–22).

The thickness of the reinforcing panels depends mainly on construction practice, because the internal forces (lateral, shear and tension) do not require, in general, a strong structure. Concrete walls with a thickness of 0.10...0.15 m with simple steel mesh reinforcement may be sufficient.

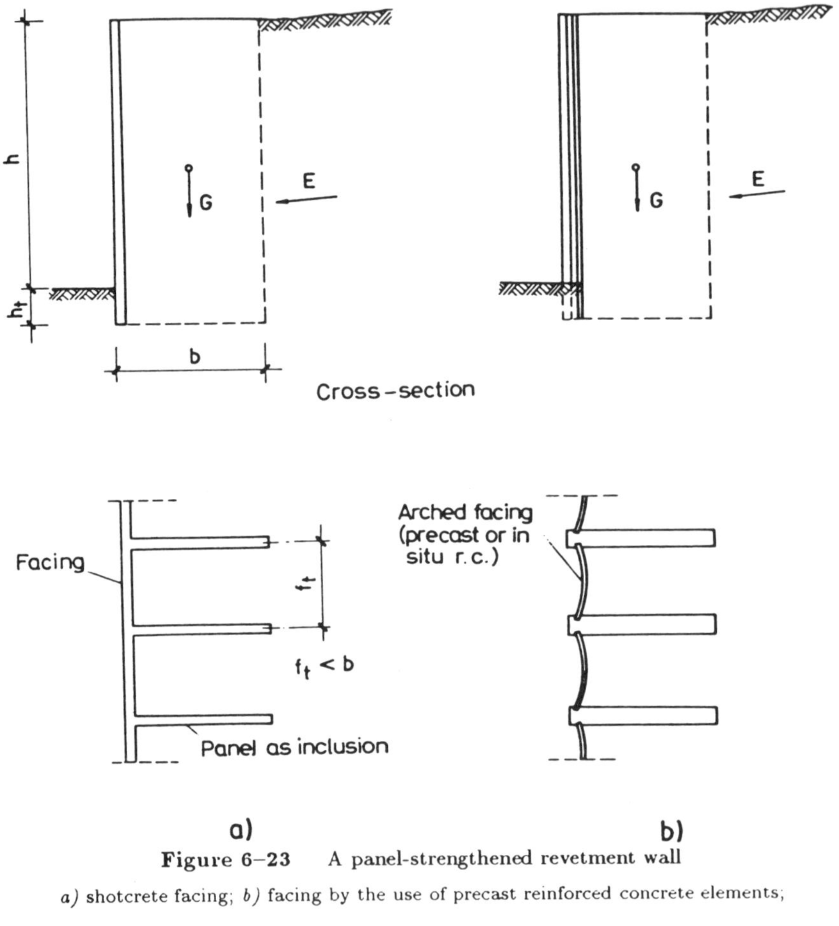

Figure 6–23 A panel-strengthened revetment wall

a) shotcrete facing; *b)* facing by the use of precast reinforced concrete elements;

As mentioned before, the construction process begins with the insertion of the panels into the soil. Different insertion technologies can be used such as trenching and pouring concrete, or precast concrete panels can be forced or vibrated into the soil. The free surface of the earth mass, in general, requires some facing (to prevent erosion or other unwanted effects). Therefore, after excavating the first earth layer of 1.0 ... 1.5 m, facing has to be installed in-between the panel sections as they are exposed. The most common solution is to place a shotcrete covering of thickness 0.06 ... 0.10 m (with steel mesh reinforcement connected to the panels), or plane or curved precast facing elements can be used (Fig. 6–23). The space behind these elements is then backfilled with granular soil or injected with cement mortar. Having finished one cycle, the next layer of soil below is treated similarly. The facing between the reinforcing panels should extend below the frost penetration zone.

The primary advantage of this type of structure lies in the minimum excavation requirement. In addition, no temporary supports are necessary. In design, the weight of the supported soil mass can be taken into account as a stabilizing term. Certainly, the construction techniques are not as common as those for some other retaining walls, and on slopes with inclinations higher than 15° this disadvantage may become serious.

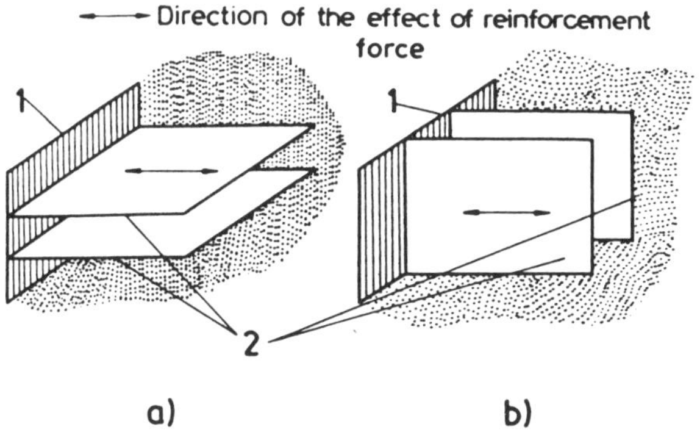

Figure 6–24 Soil reinforcement in horizontal or vertical layers. Axonometric view (*Jewell, Wishatt, 1983*)

a) horizontal reinforcement; *b)* vertical reinforcement;
1 face; *2* reinforcing layers

It is worth noting that an analogous reinforcing mechanism for walls used in offshore construction was proposed independently (*Jewell, Wishatt, 1983*). The walls of an artificial off-shore island foundation can be constructed in this manner (Fig. 6–24). Further applications of this type of construction can be found in spoil banks or waste stockpiles where the panels can be placed (and temporarily fixed) well before deposition. To determine the optimal distance between the panels and their dimensions, internal and external stability investigations are again required. Because of the novelty of these applications, we present a detailed description of them below.

6.2.2 Mechanical behaviour of panel-reinforced walls

6.2.2.1 Experimental observations

To understand the mechanical behaviour of the panel-reinforced wall, large-scale experiments were performed by the Hungarian Institute for Building Science. The dimensions of the experimental pit, which was filled with fine sand are shown in Fig. 6–25.

Narrow wooden planks of dimensions 1600×300×50 mm were pushed into the soil ($\varphi = 32°$, $c = 0$, void ratio = 0.7) at spacings of 0.3 m. The compacting effect of this process is worth mentioning. After excavating the soil in front of the row of planks, the free face of the reinforced soil wall with a height of 1.40 m was exposed. The clamped length of the planks was about 0.20 m. No facing was applied in order to make possible observation of the sliding surface. Doubly curved surfaces and a smooth logarithmic helix developed between the planks in the vertical section while shallow arches developed in the horizontal planes.

The unloaded backfill remained stable for several weeks, with only small local slips occurring between the planks. To precipitate a collapse, a surface load was applied but the wall remained stable. Diffuse localized sliding surfaces developed between the planks but the panel-reinforced soil mass moved (rotated and displaced forwards) only slightly like a composite retaining wall structure.

It is worth noting that similar failure surfaces were observed during the investigations into slurry supported trenches (*Prager, 1973*; *Karstedt, 1978, 1980*; *Walz, Prager, 1979*) and walls built with contiguous (but not interlocked) piles (*Tóth, 1978*).

The analysis of external and internal stabilities will be presented for cohesionless soils. Of course, just as for soil walls, some (at least apparent) cohesion is required in this case too to protect the face; but this requirement can easily be fulfilled. A similar approach can be used for soils with cohesion.

The reader might find it interesting to develop an advanced mechanical model which takes into account the silo and friction effects in vertical and horizontal directions, leading to a system of inhomogeneous partial differential equations. Leaving this possibility for further research, here we present a simple approximation recommended for practice.

6.2.2.2 Analysis of external stability

The interaction between the soil and the thin reinforcing panels results in a composite wall of the gravity type. Therefore, the usual failure modes considered in traditional investigations (slip, overturning, etc.) have to be considered (Fig. 6–26).

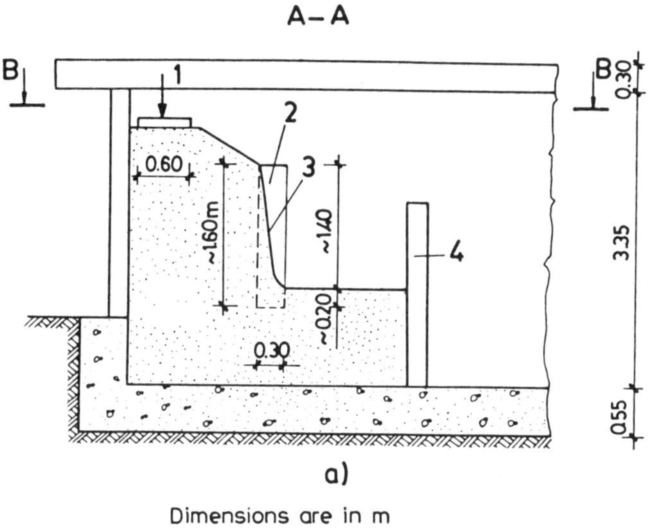

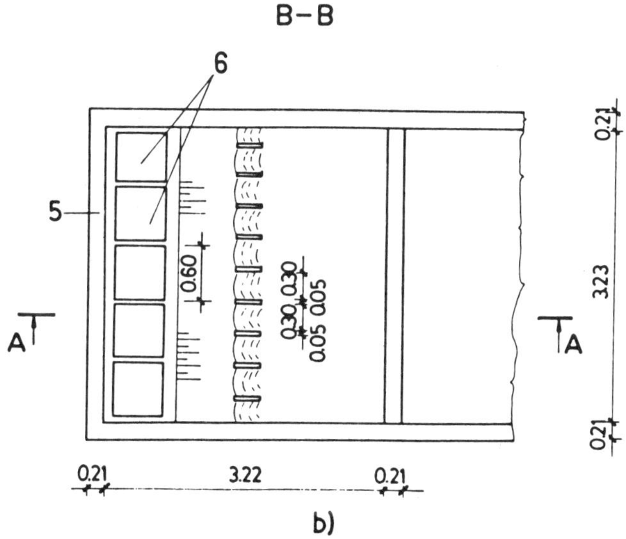

Figure 6–25 Tests on a panel revetment wall (Hungarian Institute for Building Science, 1981)

a) section; *b)* plan;

1 load; *2* timber plank; *3* sliding surface; *4* partition wall; *5* wall of the soil box; *6* loading plates

We assume cohesionless soil (the apparent cohesion, required for the temporary stages of construction can be neglected for the sake of safety). and that the direction of the active *Rankine* earth pressure acting at the back BD of the mass is horizontal. A unit width is considered in the y direction. In accordance with Fig. 6–26:

— the equilibrium condition of the horizontal forces can be expressed as $E = S$, where

$$E = \frac{h}{2}(h\gamma + 2q)\lambda_a \tag{6–12}$$

where E denotes the earth pressure and

$$S \leq (G + Q)\tan\varphi \tag{6–13}$$

represents the passive frictional force acting at the bottom of the wall; G denotes the weight of the wall and Q represents the resultant of the distributed surface loads

$$Q = qb, \tag{6–14}$$

— the equilibrium condition of the vertical forces:

$$V = G + Q,$$

— the equilibrium condition of the moments:

$$M_A = Gx_G + Q\frac{b}{2} - E(h - z_E) - Vx_V = 0. \tag{6–15}$$

The weight of the wall section (ABCD) should be calculated using the equivalent unit weight γ' (see insert at the bottom in Fig. 6–26).The negligible difference between γ' and the unit weight γ of the soil mass allows use of the latter in the calculations. Also, assuming that $x_G = b/2$, we obtain:

$$G = bh\gamma \tag{6–16}$$

and

$$E(h - z_E) = \frac{h^2}{6}(h\gamma + 3q)\lambda_a. \tag{6–17}$$

Using (6–14) and $x_V = \left(\frac{b}{2} - e\right)$ where e denotes the eccentricity of the vertical reaction force V, we obtain from (6–15):

$$e = \frac{h^2\lambda_a}{6b} \cdot \frac{h\gamma + 3q}{h\gamma + q}. \tag{6–18}$$

Introducing the notation $B_1 = \left[\frac{h\gamma+3q}{h\gamma+q}\right]^{1/2}$ we arrive at

$$e = \frac{1}{6b}(B_1 h)^2\lambda_a. \tag{6–19}$$

The soil stresses are obtained from the following expressions:
— for $e \leq \frac{b}{6}$ (Fig. 6–26a):

$$\sigma_A = \frac{V}{b}\left(1 + \frac{6e}{b}\right) \tag{6–20}$$

$$\sigma_B = \frac{V}{b}\left(1 - \frac{6e}{b}\right), \tag{6–21}$$

— for $e \geq \frac{b}{6}$ (Fig. 6–26b,c)

$$\sigma_A = \frac{2V}{3x_V}. \tag{6–22}$$

The safety against overturning can be expressed in a general form as the ratio of those moments acting for and against overturning:

$$n_f = \frac{Gx_G + Q\frac{b}{2}}{E(h - z_E)},$$

which can be rewritten (using the notation introduced above) as

$$n_f = \frac{3(h\gamma + q)}{(h\gamma + 3q)\lambda_a} \cdot \frac{b^2}{h^2} \tag{6–23}$$

or

$$n_f = \frac{3}{\lambda_a}\left(B_1\frac{b}{h}\right)^2. \tag{6–24}$$

The appropriate width b, dependent on the safety factor against overturning n_f, can be calculated from (6–23) and (6–24) as

$$b = \left(\frac{n_f(h\gamma + 3q)\lambda_a}{3(h\gamma + q)}\right)^{1/2} h \tag{6–25}$$

or

$$b = \sqrt{\frac{n_f\lambda_a}{3}}B_1 h. \tag{6–26}$$

The safety against slip, n_e will be expressed as the ratio of the active and passive horizontal force components (again, for the passive forces, their value is taken in the limit state):

$$n_e = \frac{(G + Q)\tan\varphi}{E}$$

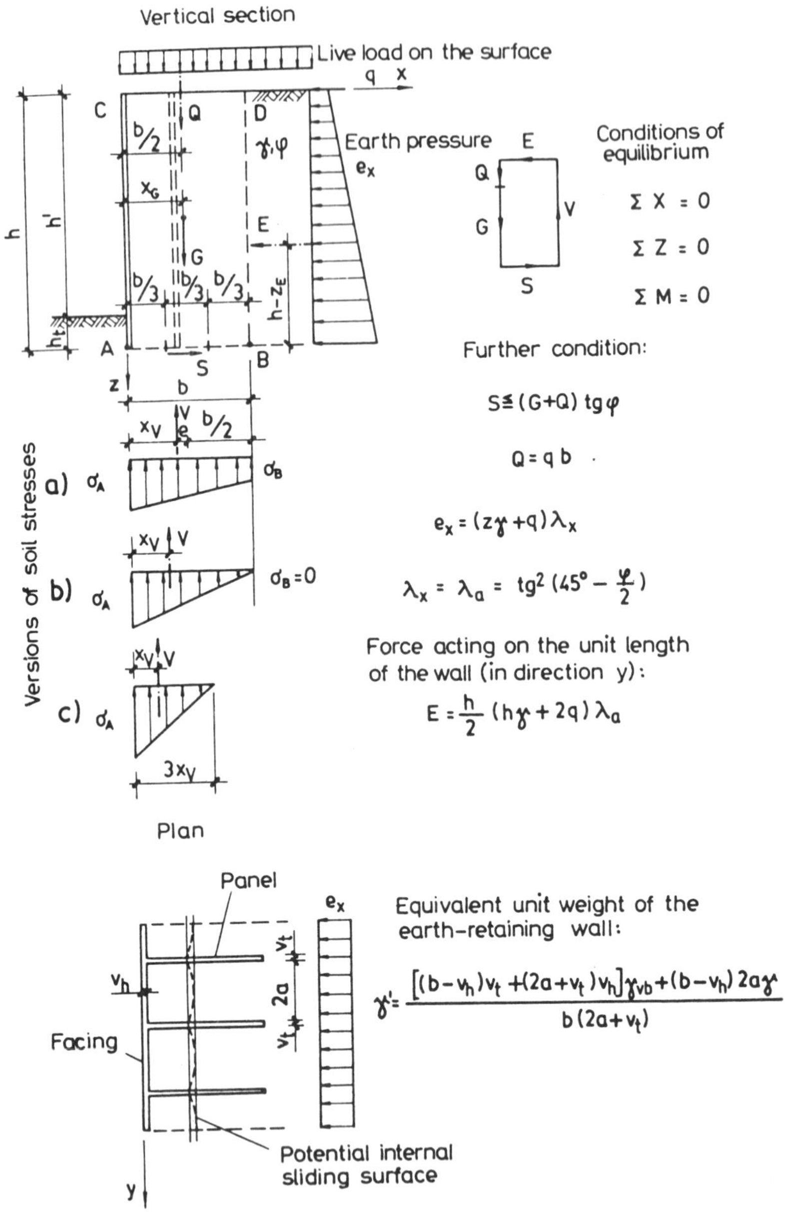

Figure 6–26 Investigation of the external stability of a panel-revetment wall

γ specific unit weight of the soil; φ the soil's angle of friction; γ_{vb} unit weight of reinforced concrete

Table 6–1 Determination of factor of safety against overturning and slip and of wall width

in investigating						
	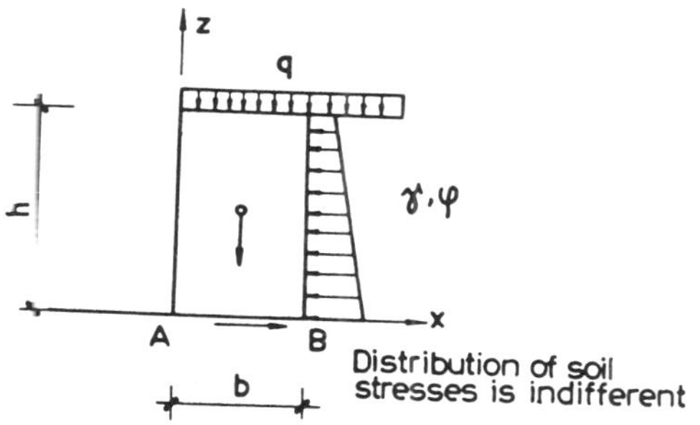		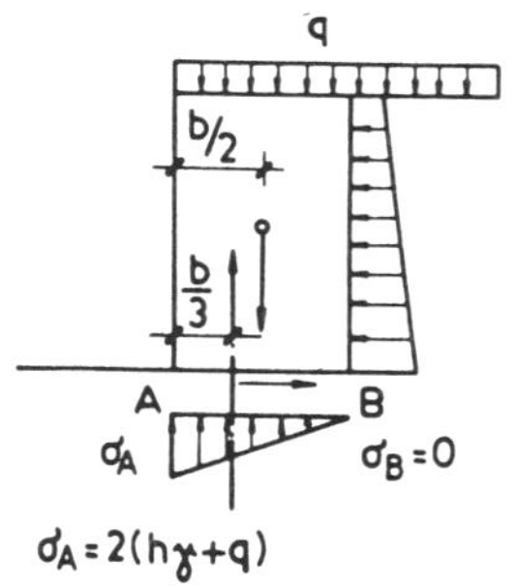		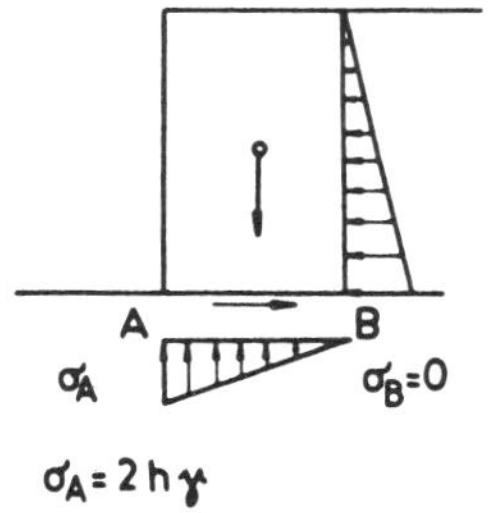	
case	General	$\varphi = 30°$	General	$\varphi = 30°$	General	$\varphi = 30°$
overturning	$n_f = \frac{3(h\gamma + q)}{(h\gamma + 3q)\lambda_a} \cdot \frac{b^2}{h^2}$	$n_f = \frac{9(h\gamma + q)}{(h\gamma + 3q)} \cdot \frac{b^2}{h^2}$	$n_f = 3.00$		$n_f = 3.00$	
overturning	$b = \left[\frac{n_f(h\gamma + 3q)\lambda_a}{3(h\gamma + q)}\right]^{1/2} h$	$b = \left[\frac{(h\gamma + 3q)}{9(h\gamma + q)}\right]^{1/2} h$	$b = \left[\frac{(h\gamma + 3q)\lambda_a}{h\gamma + q}\right]^{1/2} h$	$b = \left[\frac{h\gamma + 3q}{3(h\gamma + q)}\right]^{1/2} h$	$b = \lambda^{1/2} h$	$b = 0.577h$
slip	$n_e = \frac{2(h\gamma + q)\tan\varphi}{(h\gamma + 2q)\lambda_a} \cdot \frac{b}{h}$	$n_e = \frac{1.732(h\gamma + q)}{(h\gamma + 2q)} \cdot \frac{b}{h}$	$n_e = 2\left[1 - \frac{q^2}{(h\gamma + 2q)^2}\right]^{1/2} \frac{\tan\varphi}{\lambda_a^{1/2}}$	$n_e = 2\left[1 - \frac{q^2}{(h\gamma + 2q)^2}\right]^{1/2}$	$n_e = 2\frac{\tan\varphi}{\lambda_a^{1/2}}$	$n_e = 2.00$
slip	$b = \frac{(h\gamma + 2q)\lambda_a}{2(h\gamma + q)\tan\varphi} n_e h$	$b = \frac{0.287(h\gamma + 2q)}{(h\gamma + q)} n_e h$	The value of b was determined by conditions $\sum Z = 0$ and $\sum M = 0$; in the case of $n_f = 3.00$, $\sigma_A \neq 0$, $\sigma_B = 0$, b cannot be interpreted, respectively, it was already determined in the precedings			

n — factor of safety; $\lambda_a = \tan^2\left(45° - \frac{\varphi}{2}\right)$; $\lambda_a = 0.333$ in the case of $\varphi = 30°$

Table 6–2 Determination of major design parameters with a factor of safety against overturning $n_f = 3.00$ and with the assumption of a triangular soil stress distribution along the entire width of the wall

		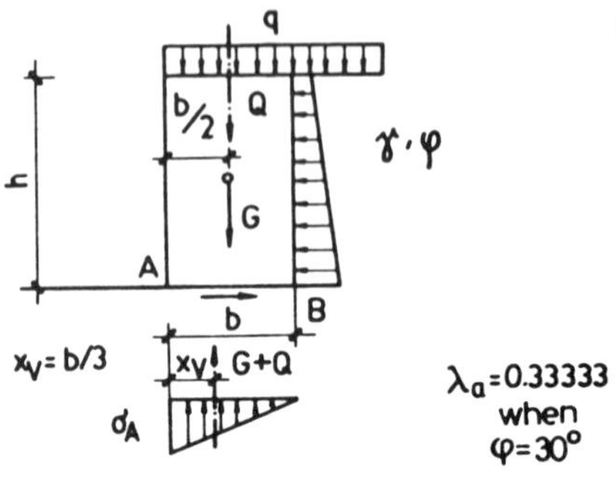		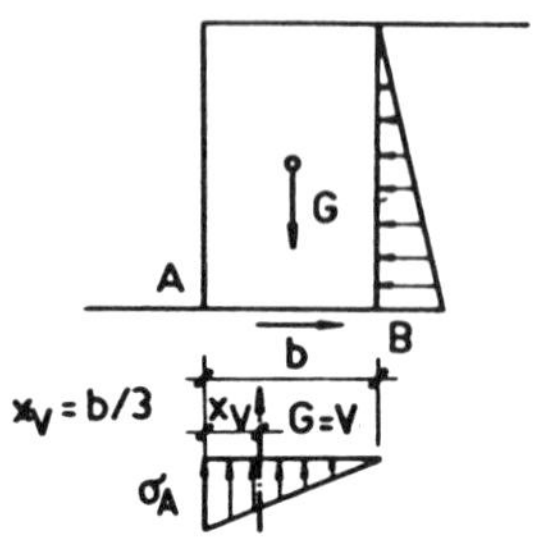	
		General	$\varphi = 30°$	General	$\varphi = 30°$
		case			
if the factor of safety against overturning is $n_f = 3.0$, then	$b =$	$\sqrt{\lambda_a} B_1 h$	$0.57735\, B_1 h$	$\sqrt{\lambda_a} \cdot h$	$0.57735\, h$
	$x_V =$	$\frac{1}{3}\sqrt{\lambda_a} B_1 h$	$0.19345\, B_1 h$	$\frac{1}{3}\sqrt{\lambda_a} h$	$0.19245\, h$
	$\sigma_A =$	$2(h\gamma + q)$		$2h\gamma$	
	$n_e =$	$\frac{\tan\varphi}{2\sqrt{\lambda_a} B_1 B_2}$	$2 B_1 B_2$	$2\frac{\tan\varphi}{\sqrt{\lambda_a}}$	2.00

n_e — factor of safety against slip; $B_1 = \left[\frac{h\gamma + 3q}{h\gamma + q}\right]^{1/2}$; $B_2 = \frac{h\gamma}{h\gamma + 2q}$;

$\lambda_a = \tan^2\left(45° - \frac{\varphi}{2}\right)$.

which can be rewritten as

$$n_e = \frac{2(h\gamma + q)\tan\varphi}{(h\gamma + 2q)\lambda_a} \cdot \frac{b}{h} \tag{6–27}$$

and the appropriate width b, depending on n_f, can be obtained in this case from

$$b = \frac{(h\gamma + 2q)\lambda_a}{2(h\gamma + q)\tan\varphi} n_e h \tag{6–28}$$

Table 6–3 Determination of major design parameters with a factor of safety against overturning $n_f < 3.0$ and with the assumption of a triangular soil stress distribution only along a part of the wall width

if the factor of safety against overturning is $n_f < 3.0$, then	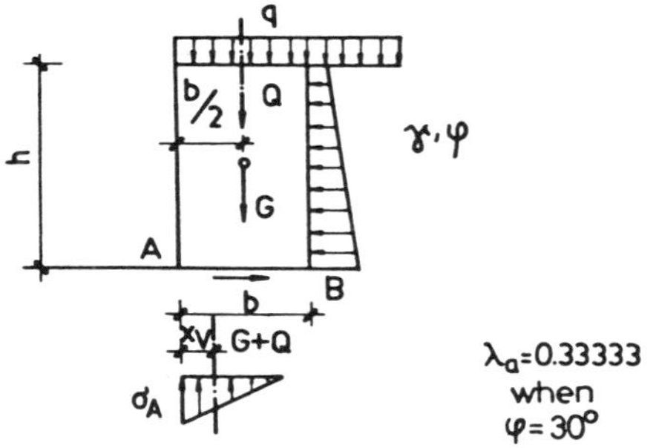		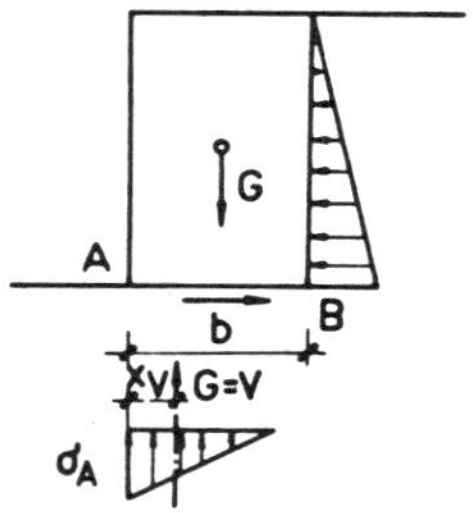	
	General	$\varphi = 30°$	General	$\varphi = 30°$
	case			
$b =$	$\sqrt{\frac{n_f \lambda_a}{3}} B_1 h$	$0.33333\sqrt{n_f} B_1 h$	$\sqrt{\frac{n_f \lambda_a}{3}} h$	$0.33333\sqrt{n_f} h$
$x_V =$	$\sqrt{\frac{\lambda_a}{3n_f}} \left(\frac{n_f - 1}{2}\right) B_1 h$	$0.16667 \left(\frac{n_f - 1}{\sqrt{n_f}}\right) B_1 h$	$\sqrt{\frac{\lambda_a}{3n_f}} \left(\frac{n_f - 1}{2}\right) h$	$0.16667 \left(\frac{n_f - 1}{\sqrt{n_f}}\right) h$
$\sigma_A =$	$\frac{4n_f}{3(n_f - 1)}(h\gamma + q)$		$\frac{4n_f}{3(n_f - 1)} h\gamma$	
$n_e =$	$2\sqrt{\frac{n_f}{3\lambda_a}} \tan\varphi B_1 B_2$	$1.15470\sqrt{n_f} B_1 B_2$	$2\sqrt{\frac{n_f}{3\lambda_a}} \tan\varphi$	$1.15470\sqrt{n_f}$

n_e — factor of safety against slip; $B_1 = \left[\frac{h\gamma + 3q}{h\gamma + q}\right]^{1/2}$; $B_2 = \frac{h\gamma}{h\gamma + 2q}$; $\lambda_a = \tan^2\left(45° - \frac{\varphi}{2}\right)$.

Table 6–4 Determination of major design parameters with a factor of safety against overturning $n_f = 2.0$ and with the assumption of a triangular soil stress distribution only along a part of the wall width

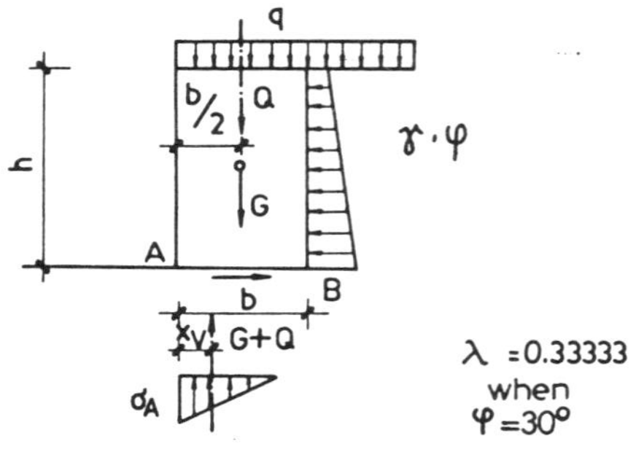

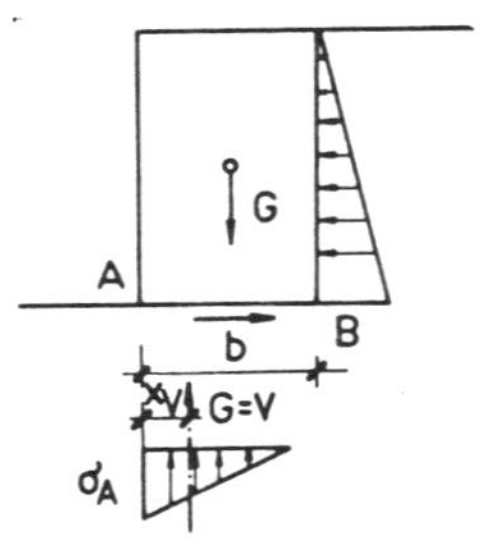

		case			
		General	$\varphi = 30°$	General	$\varphi = 30°$
if the factor of safety against overturning is $n_f = 2.0$, then	$b =$	$0.81650\sqrt{\lambda_a} B_1 h$	$0.47140\, B_1 h$	$0.81650\sqrt{\lambda_a} h$	$0.47140\, h$
	$x_V =$	$0.20412\sqrt{\lambda_a} B_1 h$	$0.11785\, B_1 h$	$0.20412\sqrt{\lambda_a} h$	$0.11785\, h$
	$\sigma_A =$	$2.66667(h\gamma + q)$		$2.66667\, h\gamma$	
	$n_e =$	$1.63299 \frac{\tan\varphi}{\sqrt{\lambda_a}} B_1 B_2$	$1.63299\, B_1 B_2$	$1.63299 \frac{\tan\varphi}{\sqrt{\lambda_a}}$	1.63299

n_e — factor of safety against slip; $B_1 = \left[\frac{h\gamma + 3q}{h\gamma + q}\right]^{1/2}$; $B_2 = \frac{h\gamma}{h\gamma + 2q}$; $\lambda_a = \tan^2\left(45° - \frac{\varphi}{2}\right)$.

Table 6–5 Determination of major design parameters with a factor of safety against overturning $n_f = 1.5$ and with the assumption of a triangular soil stress distribution only along a part of the wall width

		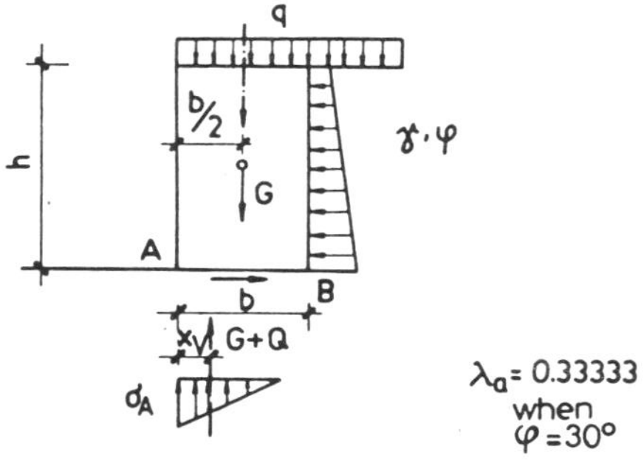		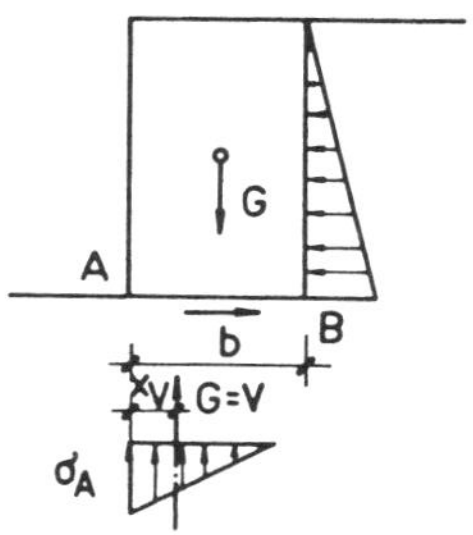	
		General	$\varphi = 30°$	General	$\varphi = 30°$
		case			
if the factor of safety against overturning is $n_f = 1.5$, then	$b =$	$0.70711\sqrt{\lambda_a} B_1 h$	$0.40825\, B_1 h$	$0.70711\sqrt{\lambda_a} h$	$0.40825\, h$
	$x_V =$	$0.11785\sqrt{\lambda_a} B_1 h$	$0.06804\, B_1 h$	$0.11785\sqrt{\lambda_a} h$	$0.06804\, h$
	$\sigma_A =$	$4(h\gamma + q)$		$4\, h\gamma$	
	$n_e =$	$1.40421 \frac{\tan\varphi}{\sqrt{\lambda_a}} B_1 B_2$	$1.41421\, B_1 B_2$	$1.41421 \frac{\tan\varphi}{\sqrt{\lambda_a}}$	1.41421

n_e — factor of safety against slip; $B_1 = \left[\frac{h\gamma + 3q}{h\gamma + q}\right]^{1/2}$; $B_2 = \frac{h\gamma}{h\gamma + 2q}$; $\lambda_a = \tan^2\left(45° - \frac{\varphi}{2}\right)$.

For the most common cases of surface loading and dimensions, Table 6–1 gives expressions for the calculation of the safety factors for the failure modes. Further relationships required for design are given in Tables 6–2...5 for safety factors against overturning.

6.2.2.3 Analysis of internal stability

We consider a panel-reinforced wall without facing and assume that active *Rankine* earth pressures act in the horizontal direction on the plane BD (Fig. 6–27). From experimental observations and theoretical considerations, we assume that lateral (horizontal) earth arches develop between the reinforcing panels supported by the frictional forces acting at the soil–panel interfaces. Earth pressure at rest can be assumed on the panel surfaces in the y direction. This value may be even higher, depending on the technique used for panel insertion. The forces acting on the wall mass between two panels are shown in Fig. 6–27.

To describe the equilibrium conditions, we consider a cell of thickness dz, with the elementary forces dE_x and dE_y, acting in the horizontal plane in the x and y directions, respectively. If μ denotes the coefficient of friction along the soil–panel interface, the frictional forces acting along the cell can be expressed as $dS = \mu\, dE_y$. We assume furthermore, that the soil cell is bounded by a parabola, and that the inclination of the soil stress resultant at the end points of this curve must be tangential (see Fig. 6–27*e*).

Using this assumption the forces acting on the elementary soil volume can be expressed as follows:

$$dE_x = e_x 2a\, dz = (z\gamma + q)\lambda_x 2a\, dz \qquad (6\text{–}29)$$

$$dE_y = e_y b_x\, dz = z\gamma\lambda_y b_x\, dz \qquad (6\text{–}30)$$

$$dS = \mu\, dE_y = \mu z\gamma\lambda_y b_x\, dz \qquad (6\text{–}31)$$

(in the expression for dE_y the effect of surface loads has been neglected for the sake of simplicity).

The equilibrium condition in the horizontal direction x can be written as

$$dE_x - 2\, dS = 0.$$

From expressions (6–29) and (6–31) we obtain

$$(\gamma z + q)\lambda_x a\, dz - \mu z\gamma\lambda_y b_x\, dz = 0 \qquad (6\text{–}32)$$

and from this equation we obtain:

$$b_x(z) = \frac{(z\gamma + q)}{z\gamma} \cdot \frac{a\lambda_x}{\mu\lambda_y}. \qquad (6\text{–}33)$$

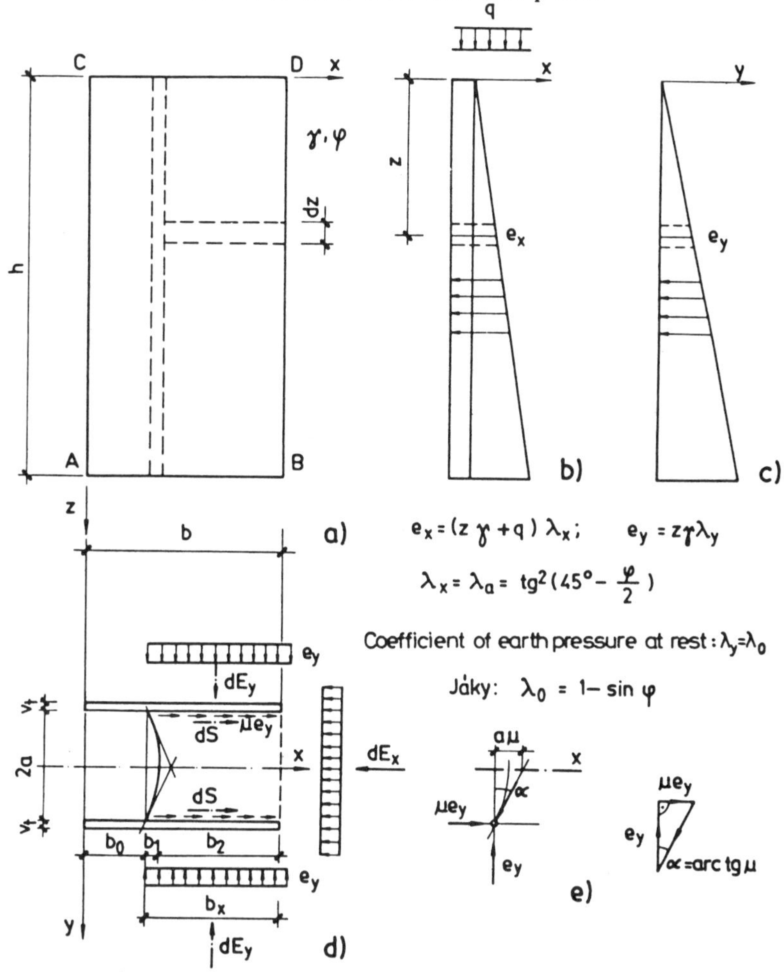

Figure 6–27 Internal stability of a panel-revetment wall

a) wall cross-section; *b)* horizontal earth pressure acting on the wall in direction x;
c) earth pressure (at rest) acting in longitudinal direction (y) on the wall;
d) plan of a wall section with external horizontal loads acting on it and showing the horizontal arches developing at failure;
e) equilibrium of stresses at the abutments

If $q = 0$,

$$b_x(z) = \frac{a\lambda_x}{\mu\lambda_y}. \tag{6–34}$$

Expression (6–33) is meaningless if $z = 0$, therefore, if $q \neq 0$, b_x must be calculated by considering the overall equilibrium condition over the whole height of the wall. Hence

$$E_x = \frac{h}{2}(h\gamma + 2q)\lambda_x 2a$$

$$E_y = \frac{h^2\gamma}{2}\lambda_y b_x$$

$$S = \mu E_y = \mu\frac{h^2\gamma}{2}\lambda_y b_x$$

$$E_x - 2S = 0$$

$$\frac{h}{2}(h\gamma + 2q)\lambda_x a - \mu\frac{h^2\gamma}{2}\lambda_y b_x = 0 \qquad (6\text{–}35)$$

and therefore

$$b_x = \frac{(h\gamma + 2q)}{h\gamma} \cdot \frac{a\lambda_x}{\mu\lambda_y}. \qquad (6\text{–}36)$$

In accordance with engineering practice, a safety factor should be introduced with respect to b_x, as well. Assuming a trial value for a, the calculated b_x must be smaller than the one obtained from the external stability analysis. In addition, a may depend on the temporary conditions imposed during the construction process. As a rule of thumb, $a < 2.0$ m is suggested. If the safety factor for b_x is n (we suggest 1.5 for this parameter), then for design purposes we obtain:

— for $q = 0$ (no surface load)

$$b_{xt} = n\frac{a\lambda_x}{\mu\lambda_y}, \qquad (6\text{–}37)$$

— for $q \neq 0$

$$b_{xt} = n\frac{(h\gamma + 2q)}{h\gamma} \cdot \frac{a\lambda_x}{\mu\lambda_y}. \qquad (6\text{–}38)$$

6.2.2.4 Design of the facing

To dėtermine the loads acting on the facing we assume a 'silo stress state' between the facing and the potential slip surface in the soil wall mass. The effect of the surface load acting along the line CD can be described (with the notation given in Fig. 6–28) by

$$e_h(z = 0) = q\lambda_a. \qquad (6\text{–}39)$$

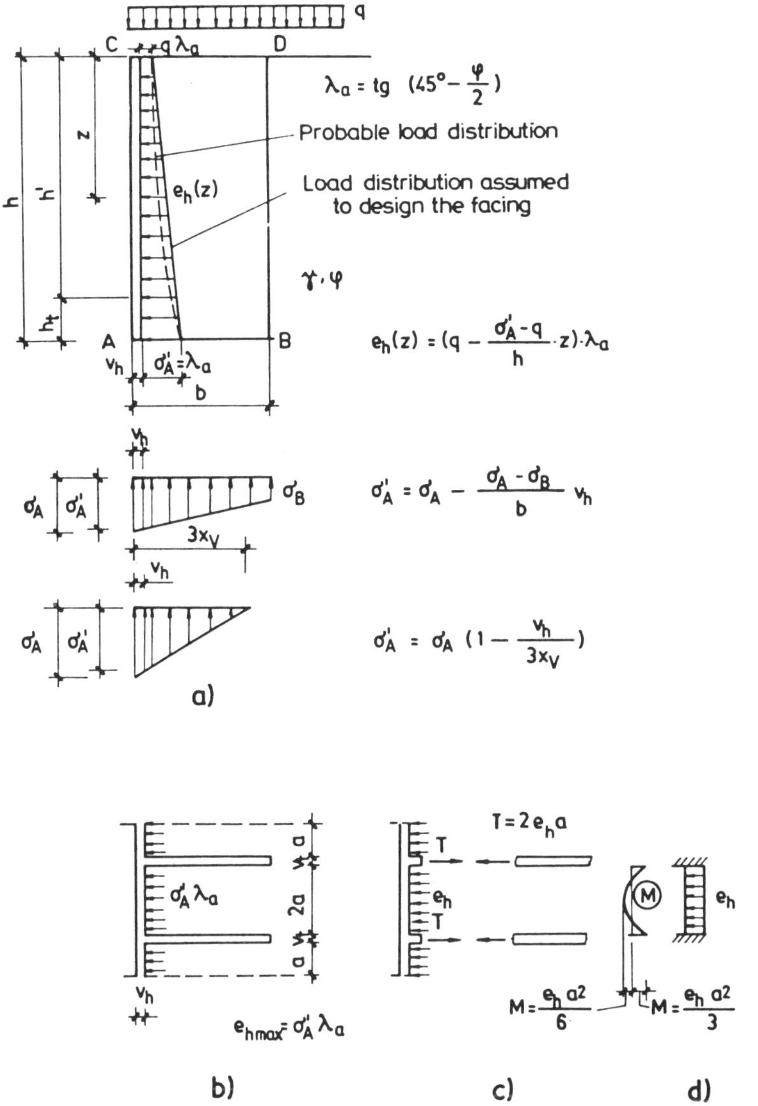

Figure 6–28 Design of the facing of a panel-revetment wall

a) cross-section; *b)* plan; *c)* forces acting on the facing;

d) moments arising in the face element of the jointed r.c. facing–panel structure

Along the boundary AB, the soil stresses are taken from the external stability analysis. Using σ'_A from (6–20) the lateral earth pressure can be written as

$$e_h(z = h) = e_{h\,max} = \sigma'_A \lambda_a \qquad (6\text{–}40)$$

For design purposes, a linear variation can be assumed between these top and the bottom values. In practice, the dimensions are usually governed by practical constructional considerations and these are usually acceptable. In general, because of the silo effects, the pressure at the facing may be even smaller than that obtained from the simplified analysis (Fig. 6–28*a*). If some structural details such as connections between precast panels and the precast facing require detailed design, the usual assumptions of engineering practice can be made. Some straightforward simplifications are shown in Fig. 6–28*c* and *d*.

6.2.3 Numerical example

To illustrate the steps in the design, we consider a simple example of a wall of height $h = 5.0$ m with a clamped section of height $t = 1.0$ m (therefore, $h' = 4.0$ m, see the figure in Table 6–6). Let the characteristic parameters of the soil be:

— $\gamma = 18$ kN/m^3, unit weight,

— $\varphi = 30°$, angle of internal friction,

— $\mu = \tan\left(\frac{2}{3}\varphi\right) = \tan 20°$,

— A pressure $q = 10$ kPa is assumed on the surface.

First, the wall width b and other major parameters must be determined for the factors of safety against overturning $n_f = 3.0$; $n_f = 2.0$; $n_f = 1.5$. The calculations of external stability are summarized in Table 6–6.

In the following, the analysis of the internal stability of a revetment wall of $b = 2.58$ m width and the design of a facing will be carried out assuming a safety factor of $n_f = 2.0$ against overturning. Assuming $a = 0.75$ m and taking $n = 1.50$, on the basis of expression (6–38) we obtain:

$$b_{xt} = 1.50\frac{(5.0 \cdot 18.0 + 2 \cdot 10.0)}{5.0 \cdot 18.0} \cdot \frac{0.75\lambda_x}{\mu\lambda_y} = 2.52 \text{ m} < b$$

what means that the design is acceptable.

From Fig. 6–28 and Table 6–6, taking $v_t = v_h = 0.15$ m we obtain:

$$\sigma'_A = 267\left(1 - \frac{0.15}{3 \cdot 0.64}\right) = 246 \text{ kN/m}^2, \qquad \text{and}$$

$$e_{h\,max} = 246\,\lambda_a = 82 \text{ kN/m}^2.$$

Table 6–6 Numerical example of a panel wall
Calculation of external stability

The parameter's		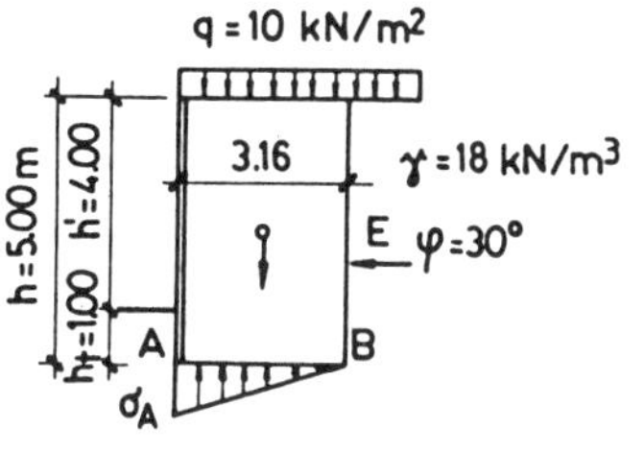	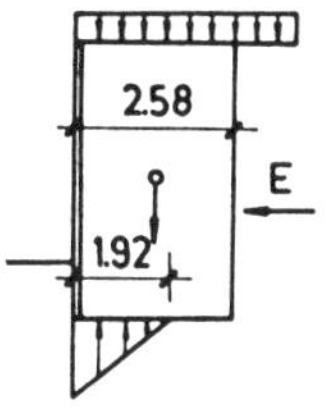	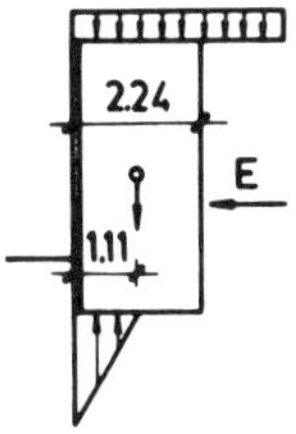
sign	unit	factor of safety against overturning: n_f		
		3.00	2.00	1:50
b	m	3.16	2.58	2.24
x_v	m	~ 1.05	0.64	0.37
σ_A	kN/m^2	200	267	400
n_e	1	1.79	1.46	1.27

n_e — factor of safety against slip

The force at the joint between the facing and the panel is:

$$T = 82.0 \cdot 2 \cdot 0.75 = 123.0 \text{ kN/m}.$$

Increasing T by a factor of safety of 1.1, the result is:

$$T_m = 1.1 \cdot 123.0 = 135.3 \text{ kN/m}.$$

The limit tensile stress in the concrete is:
$\sigma_{bh} = 1000$ kN/m^2 (10 N/mm^2) so, the tensile stress developed in the panel is:

$\sigma_t = \dfrac{135.3}{1.00 \cdot 0.15} = 902$ kN/m$^2 < \sigma_{bh}$, this means that the thickness of the panel is acceptable; no cracks will occur.

The reinforcement of the facing has to be determined for the greatest bending moment (Fig. 6–28*d*), the value of which, with a factor of safety as above is:

$$M_{max} = 1.1 \frac{82.0 \cdot 0.75^2}{3} = 16.91 \text{ kNm/m}.$$

The details of the revetment wall are shown in Fig. 6–29.

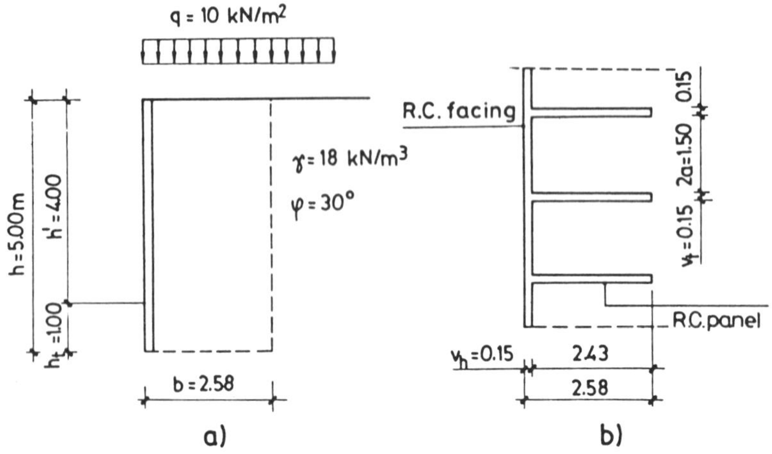

Figure 6–29 A panel-revetment wall designed for a safety factor of $n_f = 2.00$ against overturning (relevant to the numerical example in Section 6.2.3)
a) cross-section; *b)* plan

The specific volume of concrete required for the construction of a wall section of one meter length, is:

$$\frac{(b - v_h)v_t h + (2a + v_t)v_h h}{2a + v_t} =$$

$$= \frac{(2.58 - 0.15) \cdot 0.15 \cdot 5.0 + (1.5 + 0.15) \cdot 0.15 \cdot 5.0}{1.5 + 0.15} = 1.83 \text{ m}^3/\text{m}$$

which would be equivalent to a solid wall of height 5.0 m and a thickness of $1.83/5.00 = 0.37$ m. It is easy to see the economy realized in comparison with a gravity wall which would be much thicker.

7

STRUCTURAL ALTERNATIVES FOR EARTH WALLS

In the preceding six chapters, we have discussed the properties of soils, major aspects of earth structures, and the design of earth walls. Several examples were described and structural details of some basic variations have been dealt with. Some further structures of interest will be described in the following.

Any classification of earth walls will be omitted from this discussion because past experience shows that earth wall development continues to produce new hybrid solutions. Practical considerations will finally dictate the choice of structure. This choice is understood in the widest sense, taking into account, for example, local laws for industrial, environmental protection and business interests. The general use of the most efficient structures or the adoption of promising new structures may be prevented or delayed e.g. by the low rate of return of the money invested in former practices.

The functions of a retaining structure are also not appropriate for classifying them. The choice between alternatives also depends on the nature of the problem, whether it is landscape architecture, agricultural land protection, temporary or permanent regulation of water courses, or noise attenuation, etc. In poorly developed countries there may be insufficient funds to construct even small retaining walls to prevent erosion of sloping agricultural land whilst in other countries, along motorways, there may be kilometres of noise attenuating high walls built of precast reinforced concrete structures and planted by vegetation.

For these reasons, the principal types of earth-wall structure, separable by their function and structural features, are discussed below in no particular order. The design methods, which were discussed in the previous chapters, are not repeated here.

7.1 CRIB AND BOX WALLS

Local on-site soils were used for a long period of development of earth structures, to increase the mass of retaining walls. Instead of solid concrete or stone walls hollow segmental structures were built, the inside of which were then filled with earth. In this way, crib and box walls developed.

From the point of view of earth reinforcing, these retaining walls are of marginal importance as the strength of the soils is barely exploited. It is worth mentioning that the increased mass of the earth fill results in a favourable distribution of the internal forces and reduces costs. A short review of these walls is now given because they reflect the course of development of engineering techniques and, as will be seen, the idea of soil reinforcing appears in these constructions.

7.1.1 Crib walls

Stacked crib walls, made of precast reinforced concrete elements, have been known for decades in practice and their design has appeared in several publications (e.g. *Kovácsházy, 1966*). The general layout of such walls is shown in Fig. 7–1. It consists of stacked hollow prefabricated elements (wood, or r.c. members) which are then filled with earth.

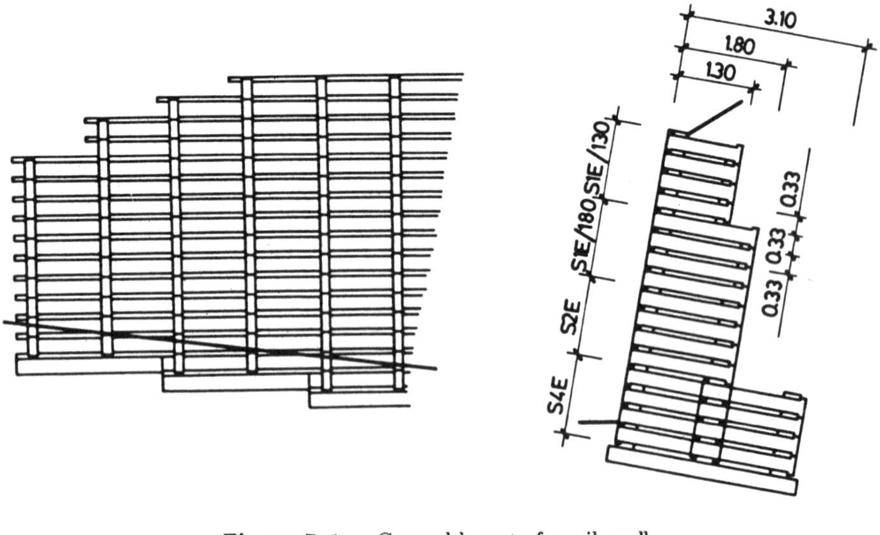

Figure 7–1 General layout of a crib wall

In the structural design the following factors should be taken into account:

— the load acting on a vertical or battered (up to 1 in 8) face of a crib wall differs from the loading pattern on gravity walls because on the free face the silo pressure acts in the internal earth core (Section 1.3). The back wall of the retained earth mass is subjected to the difference between the earth pressure and the silo pressure. The closed end faces, in the direction normal to the wall, bear the silo pressure,

— in gently inclined walls (batter less than 1 in 8) the loading is very complicated and no general theoretical model has been produced for calculation purposes,

— in checking stability, relative displacement between the earth fill and the crib wall structure should be taken into consideration: if the bottom of the structure is not closed, then in calculating the weight of the earth core, only that part of the weight, which arises from silo pressure and the interaction through friction of the wall structure and the filling material should be taken into account.

Generally, the construction of these walls does not require any special skill and they can be completed quickly. Displacements are smaller if earth core filling and stacking takes place simultaneously. Careful compaction is advantageous since the interaction between the earth core and the stack will be increased. There are two further advantages in favour of these structures: generally, dewatering can be simply carried out and, by planting the face with vegetation, the wall achieves an aesthetic appearance which blends in with the natural environment.

The structural layout of these walls has, of course, many variations. Nearly all railway companies and precast concrete plants have their own types of element—many of these can satisfy special geometric and aesthetic requirements. Recent examples (Fig. 7–2; *Brandl, 1981*) show that these structures, combined with anchorages, can also be very effective.

The advent of crib walls resulted in significant economies in the nineteen-fifties. The material requirements of the common types (walls composed of r.c. elements) in relation to wall face area is about 0.5 m^3/m^2. From cost analysis based on the price conditions of the mid sixties, this type of wall was the cheapest among those known at that time: its cost was only two-thirds of an equivalent gravity wall.

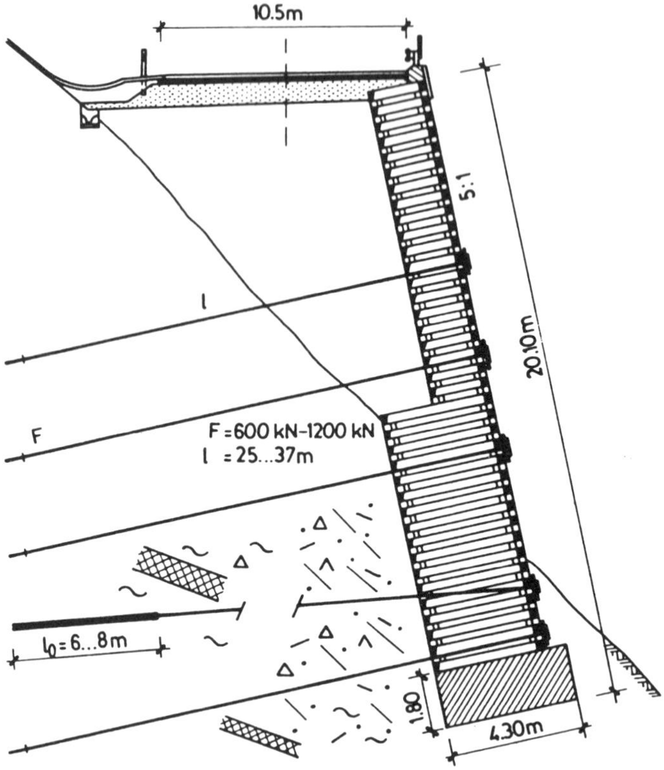

Figure 7–2 Section of an anchored crib wall (*Brandl, 1981*)

7.1.2 Closed box walls

Closed box walls are gravity walls consisting of precast elements. The mass of the earth used for filling can be exploited.

This type of structure has a long history and rich technical literature. A recent example, a Polish proposal, is shown in Fig. 7–3 (*Jarominiak, 1982*) and a licenced version thereof (*Kurtz, 1977*) in Fig. 7–4.

Structural design of the elements and detailed analysis can result in an economic structure but only if careful economical analyses justify it. Care has to be taken to prevent soaking and freezing of these structures and to provide good dewatering systems. These details are beyond the scope of our book and the reader is advised to gain further information from the technical literature (*Kovácsházy, Balázsy, 1986*). The only purpose of mentioning closed box walls here is that they can be combined with a layered soil reinforcing method and thus, their economy can be enhanced.

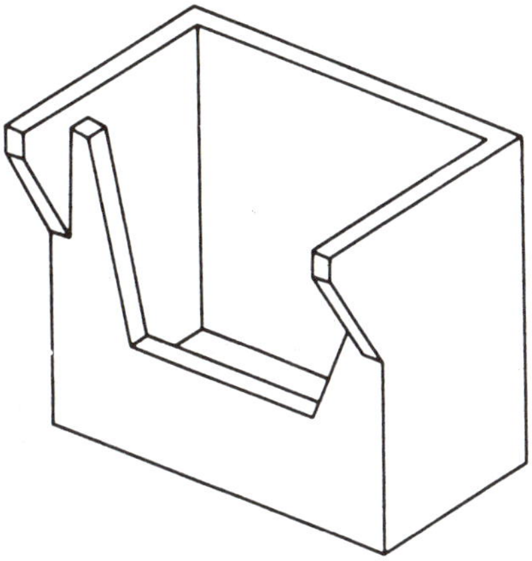

Figure 7–3 Simple closed box wall element (*Jarominiak, 1982*)

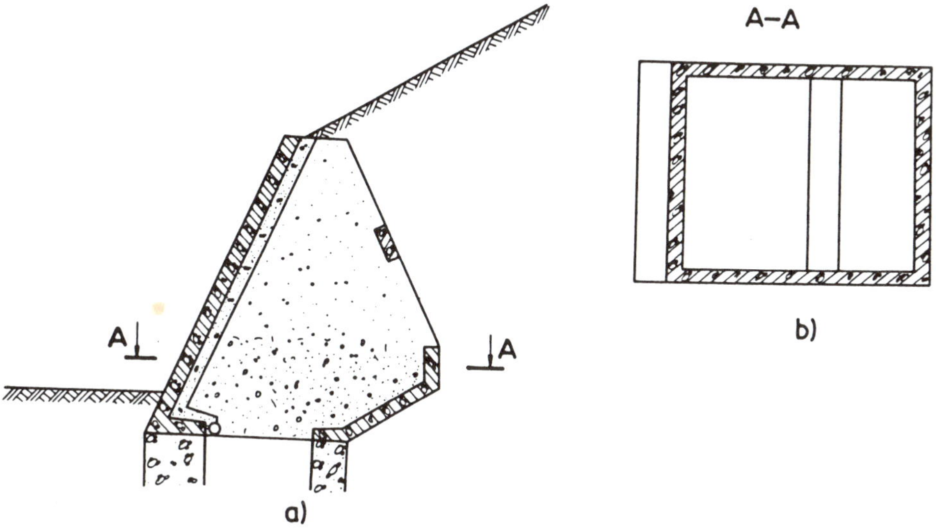

Figure 7–4 Closed element for an inclined box wall (*Kurtz, 1977*)
a) plan; *b)* section A–A

7.1.3 Open box walls

From the point of economy, open box walls are competitive and enjoy frequent application even now. A typical example is shown in Fig. 7–5. This structure was constructed at the end of the 1970's in Austria. The

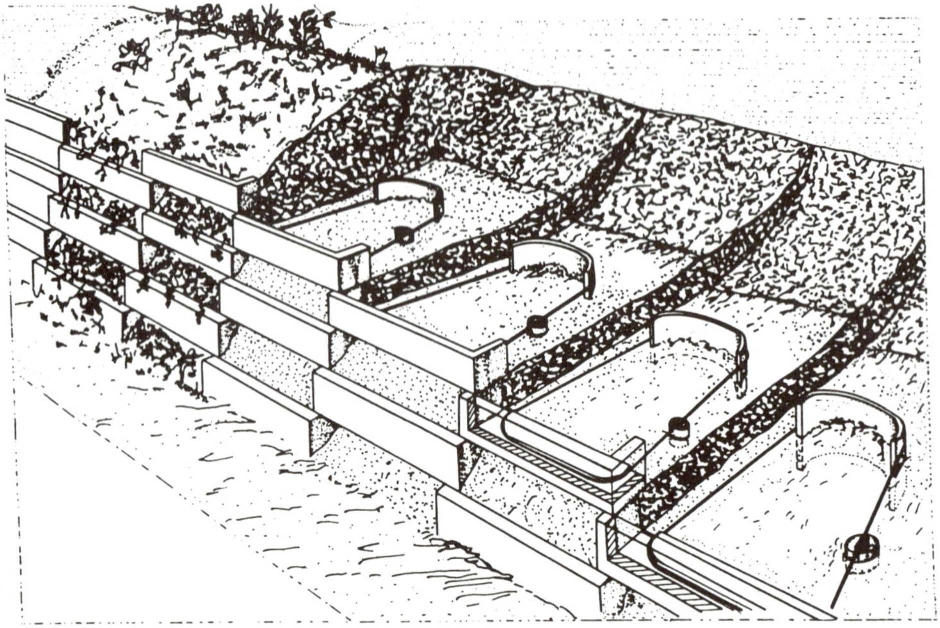

Figure 7–5 General layout of an open box wall (NEW)

sizes (Fig. 7–6) of the precast reinforced concrete elements demand mechanized construction techniques.A face area of 1.65 m^2 is covered by a single element (NEW).

The face elements are laid stepwise in a chess-board pattern. They are held in place by a single curved endless strap which is fixed to the sides of the flat bank standing out from the horizontal slab of the face element. The strap encloses the arched anchoring element embedded into the earth mass to be retained. The distance between the face and anchor element depends on the external dimensions of the wall; that is generally 2...3 m.

The average slope of the face can be changed between relatively wide limits by placing the superincumbent elements as desired. In the process of design—analysis of external stability—the tensile stresses in the straps must be checked. On the basis of experimental experience and theoretical considerations, the factor λ_a (active earth pressure) can be used, but it is advisable to calculate the tensile forces in the strap with the factor λ_0 (earth pressure at rest).

The following sequence of construction is suggested:

— preparation of a footing layer 1 m wide of lean concrete or compacted sandy gravel,

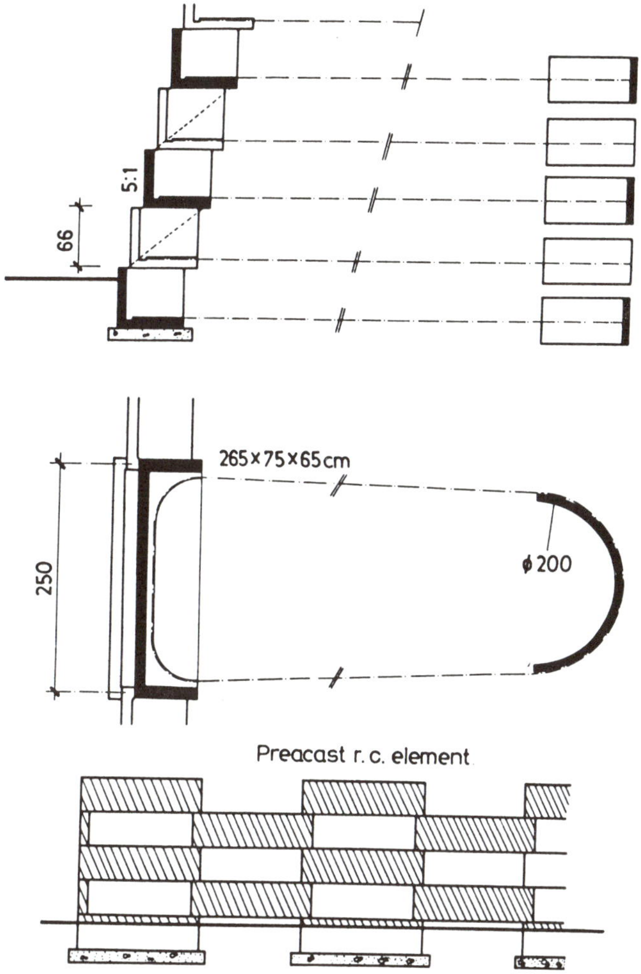

Figure 7–6 Open box wall. Dimensions and elevation (NEW)

Figure 7–7 Noise attenuating earth retaining wall beside a motorway (*dabau, 1976*)

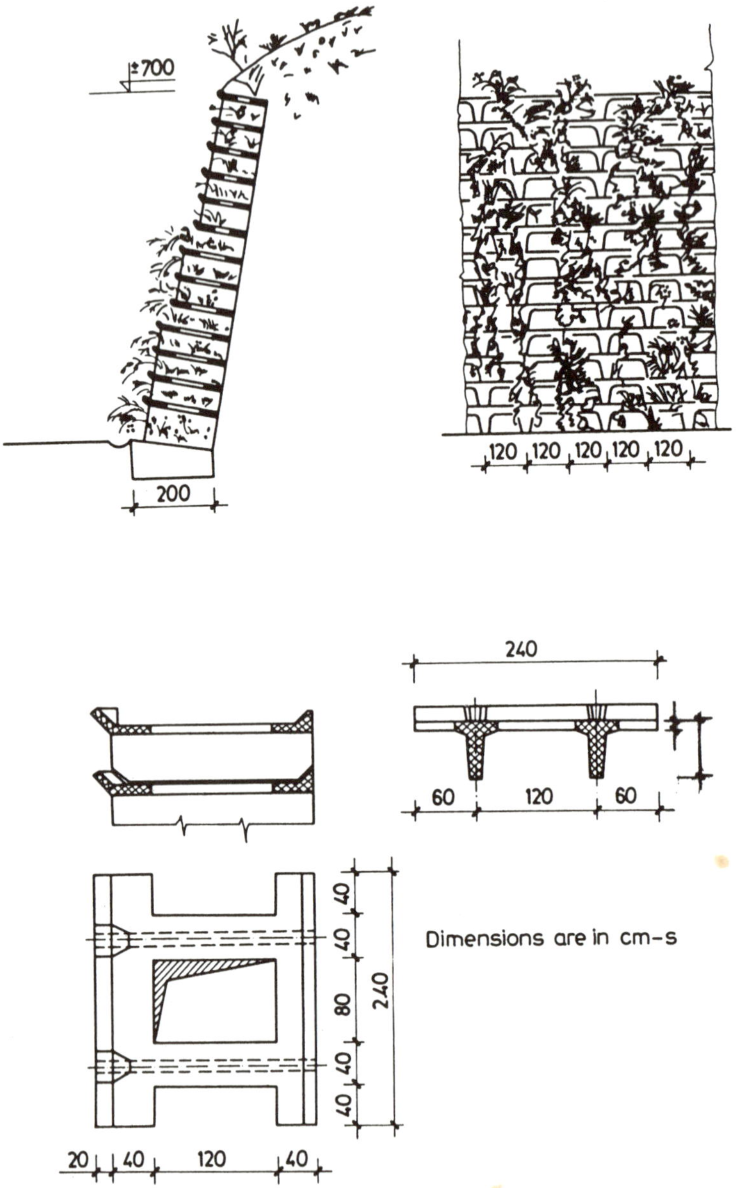

Figure 7–8 Box wall consisting of prefabricated reinforced concrete elements (*Polgár, 1983*)

— placing the face and anchor elements in the required position with respect to the anchoring strap,
— tailoring the anchoring strap to the site and placing and fixing it,
— spreading a layer of the filling material,
— compacting the fill (started from a distance of 1 m behind the face element and advancing toward the anchor element),
— surface finishing near to the face element.

Using this sequence of construction, subsequent movements will be minimized.

Planting vegetation into the slots among the face elements is worthwhile, of course; it is not only advantageous aesthetically, but experience shows that the erosive effects of surface and seepage water can be prevented in this way.

In countries with highly developed horticultural industries several contractors manufacture smaller but similar box-elements which can be used to build such structures. Fig. 7–7 shows a sketch of a noise attenuating wall (*dabau, 1976*). The reader may find many similar sketches in the advertisements of technical periodicals.

It is worth mentioning an example from Hungary where prefabricated and marketed reinforced concrete elements (originally designed for floor construction) has been used for retaining wall construction. By introducing some modifications, these elements are now suitable for constructing crib- and open box walls (*Polgár, 1983*; in Fig. 7–8). Retaining the cross section of the joists for the floor construction, shorter (length of 1.5...3.0 m) elements can be used. The hole in the horizontal slab is advantageous for the following reasons: the weight (and costs) decrease and earth filling is simpler. The sloped end of the element enhances the growth of vegetation planted into the earth fill.

7.2 LIGHTWEIGHT RETAINING WALLS FOR GARDENS

In recreational areas and gardens, land utilization rarely allows sloping surfaces. For gardens and playing grounds, it is often reasonable to modify ground conditions by simple light-weight structures.

Some publications (*Balogh, 1976*; *Pirk, 1977*; *Lelkes, 1982*) review the most important traditional solutions, which commonly feature the almost exclusive use of natural building materials (wood, stone). Where sufficient quantities are available, these materials may be preferred. If well constructed, they are stable and have an appealing appearance which blends into the landscape.

The situation changes where natural materials are surcharged by purchase and transport costs. In such cases, earth materials come into their own and only the facing will be artificial.

One sign of high level horticultural activity is the widespread use of prefabricated elements for these purposes. Usually gravity walls or box walls make use of small size concrete elements. The majority of these prefabricated building systems are patented and the technical details are not made public. The popularity of these wall types is mainly due to fashion and to competition between contractors.

Some examples of simple wall types will be reviewed in the following; these are applicable not only for residential areas but also for gardens in hilly countries.

7.2.1 Framed small retaining walls

A wall system described by the Civil Engineering Research Institute in Bratislava (*Petrik, 1975*) is assembled from simple prefabricated reinforced concrete plates arranged to form a frame (Fig. 7–9). The heavier horizontal plates weigh not more than 60 kg. They are sufficiently wide to contain earth within the frames (at a natural slope). In this way the mass and stability of the structure is increased. The stability of the structure is ensured by anchorages fixed to the elements and embedded in the backfill. Fixing of the anchors (wires, strips or rods) and shaping of their free ends have numerous variants, some of them patented.

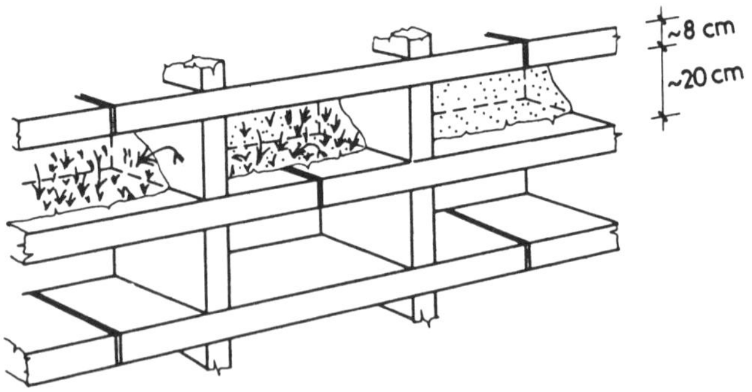

Figure 7–9 Axonometric view of a small retaining wall composed of reinforced concrete plates to form a frame (*Petrik, 1975*)

The pleasing aesthetic sight of the wall deserves attention: the free earth faces in the frames can be planted with vegetation. The material requirements what is characterized by the ratio of the volume of concrete elements to the face area is in this case about 0.07...0.08 m^3/m^2. The construction is simple, only 2 or 3 workmen are needed. Water seeping into the backfill can easily flow out through the face, but erosion is prevented by the vegetation.

7.2.2 Garden walls constructed from concrete boards

As a byproduct of research in the Hungarian Institute for Building Science (ÉTI) related to reinforced earth retaining walls—reviewed in Chapter 5—a simple garden wall was designed and constructed. This structure maintains the advantages of the large size walls but it can be adjusted to the specific site conditions (*Radnóti, Scharle, Szalatkay, 1980*).

The structural layout is shown in Fig. 7–10. The size of the face elements is 30 × 60 × 6 cm and the mass of a single element is 26 kg. The reinforcing band is made of commercially available plastic which is connected to the steel reinforcement. For low walls (0.6...1.2 m), the length of the reinforcing bands may be equal to the clear height of the wall.

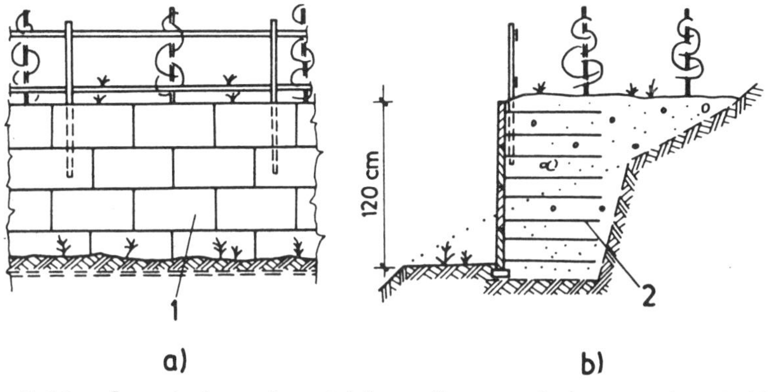

Figure 7–10 Layout of a garden retaining wall composed of concrete boards (*Radnóti, Scharle, Szalatkay, 1980*)

a) elevation; *b)* section

1 face element 30 × 60 × 6 cm; *2* plastic band

Among the advantages of concrete board walls the following should be stressed:

— the elements can be fixed by hand,

— precasting of the elements is simple and reinforcing bands are readily available,

— demand on space is small and vegetation can be planted on the wall,

— in case of damage caused by overload or erosion, the elements can be simply repaired or replaced.

Material requirements depend on the thickness of the plates but it is usally in the range of 0.05...0.06 m^3/m^2.

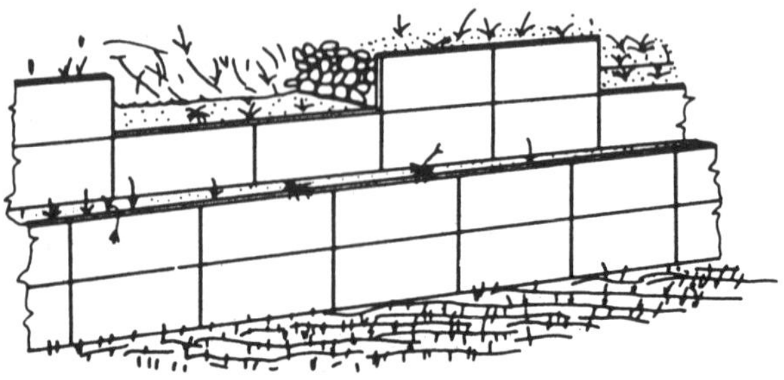

Figure 7–11 Axonometric view of the garden retaining wall in Fig. 7–10

High wall faces are sometimes overwhelming but this can be solved by building staggered walls and incorporating earth strips with vegetation. A wall built in Hungary is shown in Fig. 7–11. Instead of a smooth concrete face, it is possible to use better materials or to produce an ornamental surface finish.

7.2.3 Swedish earth retaining wall

The simultaneous use of layered reinforcement and anchorages result in a somewhat complicated but extremely safe structure as shown in Fig. 7–12. As a platform construction at a railway station, this arrangement of Swedish origin was patented in 1976 in Great Britain (*Fordervävnader AB*).

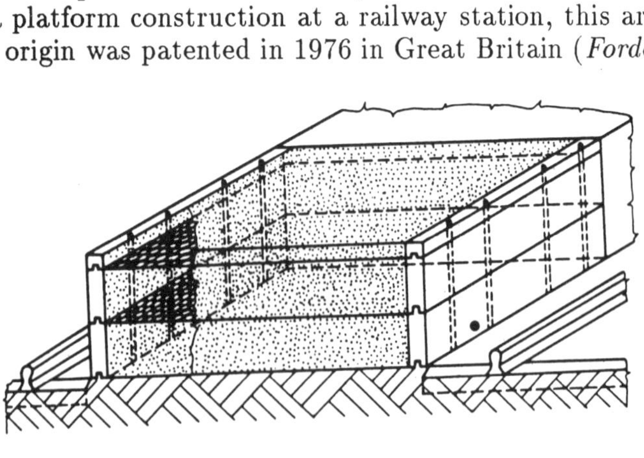

Figure 7–12 Swedish earth retaining wall; platform between railway tracks

The structure is characterized by the multiple mesh reinforcement covering the soil layers on the surface, and the fixing of the tongued and grooved face elements. The platform structure is reminiscent of an earth cofferdam.

A variant of this type of retaining wall is shown in Fig. 7–13. Having read Chapters 3 and 5, the reader may ask whether there is need for anchoring elements or strips at the end of the inclusions so close to the original ground surface. Anchoring the reinforcing meshes in this way appears to us superfluous and expensive.

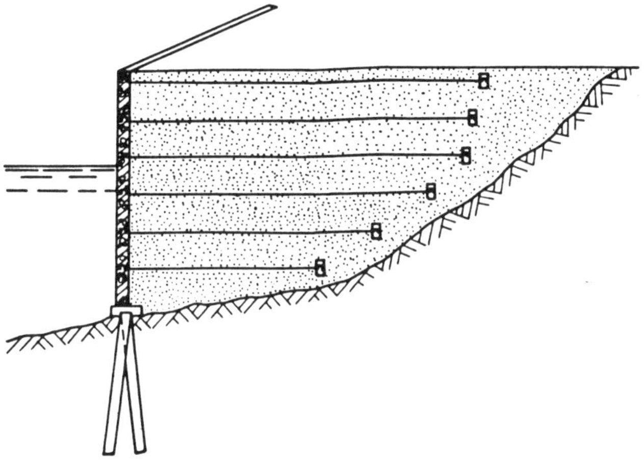

Figure 7–13 Swedish earth retaining wall; open jetty

7.2.4 Shelved small retaining wall

Walls constructed of precast elements, more so than cantilever walls, make use of the favourable effect of the earth mass acting on the backward extending horizontal structural elements. Connecting this idea with the principle of layered soil reinforcement, a further development of this wall type emerged. One such wall is shown in Fig. 7–14 (*Broms, 1977*). The reinforcing layers could be fixed to the face elements but friction on the horizontal surface would be also enough for the layers and the element to act together.

The distance between the elements (and thus that of the reinforcing layers) can be determined by combining several criteria. The design is affected by:

— the permissible weight of the precast face element,

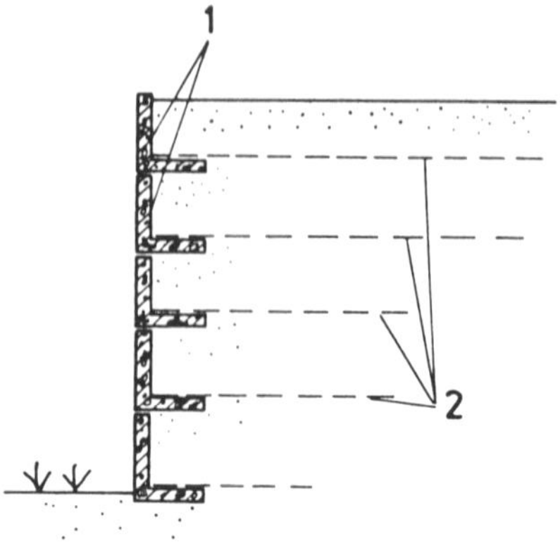

Figure 7–14 Shelved small retaining wall with layered reinforcement (*Broms, 1977*)
1 face element; *2* plastic reinforcement layer

— the thickness of the soil layers used for the backfill for effective compaction,
— the load bearing capacity and sizes of the reinforcing layers.

The basic principles of the design have been reviewed in detail in Chapter 5; approximations, of course, can be applied in simple cases. Examples can be found where an approximate distribution of earth pressure was assumed (the errors are on the safe side) and the reinforcing layers were often designed according to anchoring principles.

We have no data to estimate the economy of this wall type. However, the reinforcement in the horizontal slabs is excessive. This suggests that costs will be higher.

7.3 TEMPORARY EARTH WALLS OF LIMITED IMPORTANCE

In various ways, we have repeated the principle that the design of earth structures should reflect their importance and their benefits or losses, in case of an eventual failure of the structure. The discussion of wall types described in this section reflects these factors. The common features of these structures are as follows:
— they satisfy limited requirements both in respect of function, aesthetics and service life,
— they can be built quickly, using simple materials and construction requires less care or workmanship than the average,

— the structure must endure minor hidden inadequacies,
— their reconstruction in case of decay or damage involves costs comparable to those invested in the construction of new structures.

These features characterize the demands of landscape architecture, erosion control, agricultural land protection and related activities. For this reason in the following a review will be given of retaining wall variants applicable to this field.

7.3.1 Earth walls reinforced by using geotextiles

The theoretical and practical problems of layered soil reinforcement have been dealt with in detail in Chapter 3. Construction of earth walls of this type have become generally accepted in the past decade. They can be ideally applied for the construction of temporary or less important earthworks and they are cheap with respect to skilled labour. Their service life can be planned, their performance risk can be calculated.

We cannot review here the innumerable examples published in the technical literature. At the end of the 1970's many new net, web and sheet products appeared on the market and many publications described (*WC-80, 1980*) how they should be used in 8 ... 10 m high earth walls.

In Hungary there are many commercial products which can be used for constructing layered earth walls. The simplicity of these structures renders design almost unnecessary in the case of lower (less than 2 ... 3 m) walls. Nevertheless, it was worthwhile to check the behaviour and mode of failure of these walls by experiments because the cost of the reinforcing material is not negligible, and overdesign should be avoided.

Research to study the mechanics of reinforced earth structures with the intention of producing design rules was carried out in the Hungarian Institute for Building Science (ÉTI) with small earth retaining walls reinforced using geotextiles (Fig. 7–15).

The results can be summarized as follows (*Szalatkay, Vas, 1983*):
— The geotextile reinforced earth wall does not lose its stability under its own weight. Increasing surface load creates sliding surfaces within the reinforced zone; the most marked failures develop generally at the interface between the geotextile sheet and the retained earth mass (of course, this general form can be significantly altered under a partial surface load).
— The wall is able to bear large deformations—up to several percent—without suffering failure in the traditional sense. This deformation is not elastic. Only in the case of very short embedded geotextiles ($l/h < 0.15$) did collapse occur (l is the horizontal embedded length

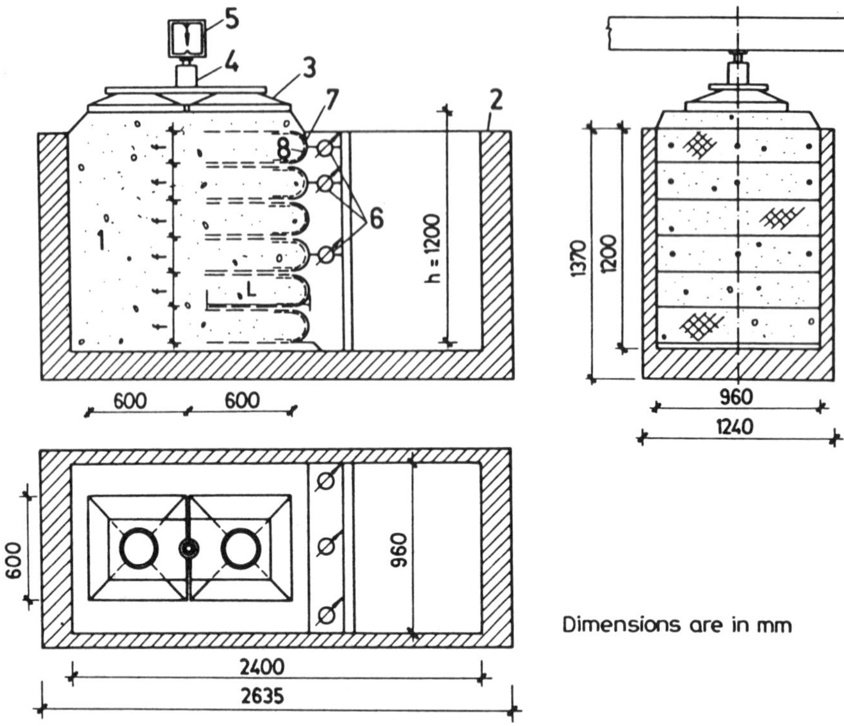

Figure 7–15 Experimental testing of an earth wall reinforced with a plastic (NETLON) web

1 soil; *2* rigid box; *3* loading plate; *4* hydraulic jack; *5* loading frame; *6* displacement meter; *7* NETLON mesh reinforcement; *8* face web; *f* vertical spacing; *l* wall width

of the geotextile, theoretically the width of the wall; h is the wall height).

- The face generally does not remain plane; on increasing the load, the stiffness of the structure reduces.
- In the case of small loads and consequently small deformation, the stiffness of the structure depends on the number of geotextile layers embedded in it, almost independently of the situation that the layers' dimension l and the spacing are large, or relative short but densely spaced geotextile layers are used.
- The bearing capacities of these retaining structures depend considerably on the properties of the soil used for construction, mainly on the angle of friction.
- The character of the relationship between loads and deformations is shown in Fig. 7–16. Although for the experiments only one type of

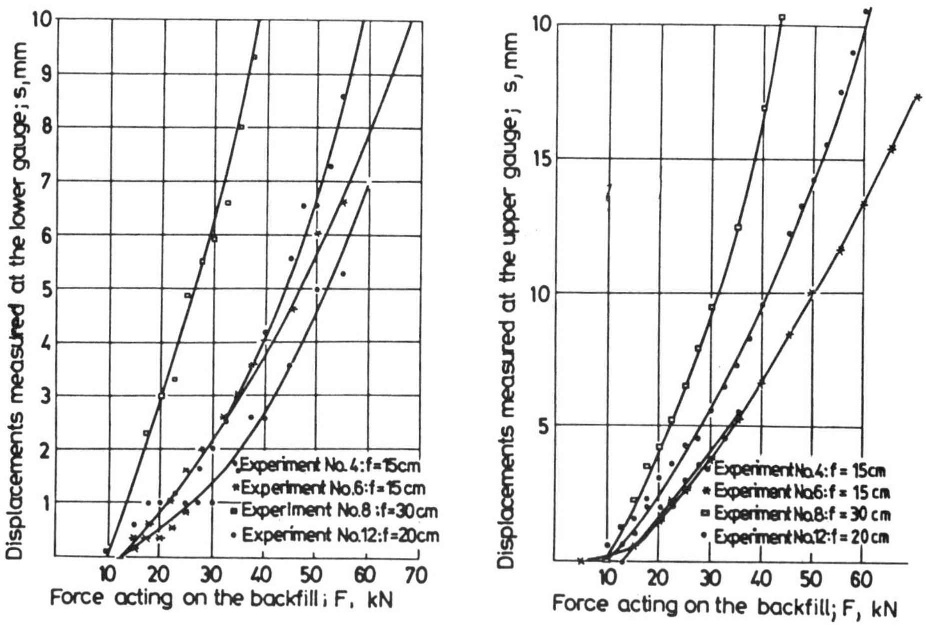

Figure 7–16 Displacement of the experimental wall's face plotted against load (diagrams of the model with l = 45 cm)

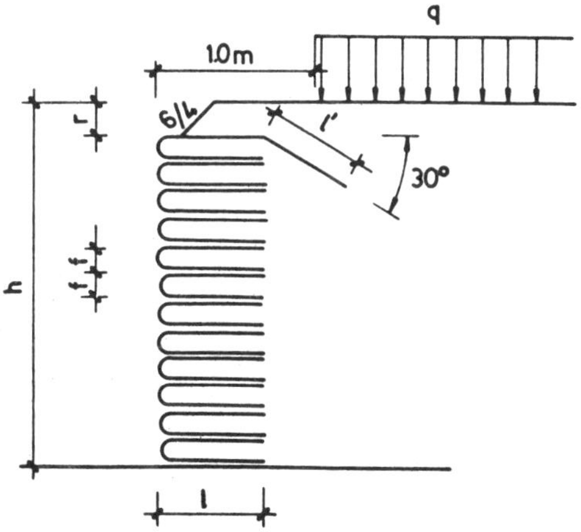

Figure 7–17 Layout of a retaining wall reinforced with a plastic (NETLON) web
γ volume weight; φ angle of friction; q surface load; l' extension; r overburden; f spacing; l width; h design height

reinforcing material (NETLON web H–12) was used, it is known from other published results that all materials suitable for reinforcement behave similarly.

The results from these model experiments made it possible to work out an approximate design method applicable to earth retaining walls reinforced with soil stabilizing NETLON webs of types H–11 and H–12. The method is shown as an example, noting that the assumptions and approximations are crude in many respects and, as may be expected, construction experience will lead to their refinement.

Data and notation are shown in Fig. 7–17. The design process will be demonstrated using the numerical data of the following example.

The design height of the retaining wall is: $h = 2.0$ m.

A uniformly distributed load (commencing at 1.0 m distance behind the face) acts on the surface:

$$q = 3 \text{ kN/m}^2.$$

For constructing the retaining wall, the local well-graded silty sand was used. Its density should reach a value of 20 kN/m^3 after compaction. The angle of friction from Table 7–1 is $\varphi = 30°$.

The permissible inclination Δ of the wall after construction is defined in Fig. 7–18. Its value is taken at present as $\Delta = 1/200$ (we shall see that the choice of inclination value may be limited by h, q and the soil characteristics).

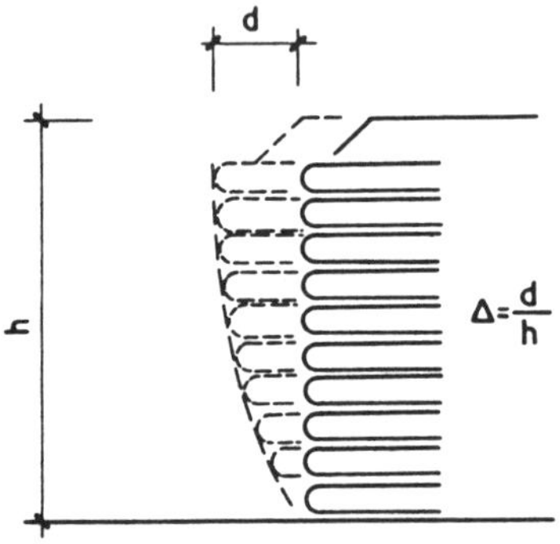

Figure 7–18 Explanation of parameters characterizing the displacement of retaining walls

Design can be carried out using Table 7–2. The auxiliary quantities found in it should be correlated and checked.

The first step in the analysis is to calculate the 'wall-height', (only to find the suitable entry in the table):

$$h^{**} = n_g h + n_q \frac{q}{\gamma},$$

Table 7–1 Characteristics of soils suitable for constructing a geotextile-reinforced retaining wall

Sort of soil	Weight per unit volume[1] kN/m^3	Angle of friction, degree
Coarse to fine gravel	19	37.5
Sandy gravel without silt, poorly graded	20	35
Coarse and medium size uniformly graded sand	18.5	33.5
Poorly graded silty sand[2]	20	30
Fine sand without silt, uniformly graded	17	32

[1] After construction, relating to the compacted state.
[2] Not to use for walls permanently immersed in water or exposed to long-lasting drenching

Table 7–2 Relationships between characteristic data of a geotextile-reinforced wall

1	2	3	4	5	6	7
	φ^o			Δ		
40	35	30	1/400	1/200	1/100	1/70
	l/h			h^{**}		
0.40	0.45	0.50	1.8...2.0	2.0...2.3	2.4...3.2	2.8...4.2
0.30	0.35	0.40	1.7...1.9	1.9...2.2	2.2...2.7	2.4...3.0
0.20	0.25	0.30	1.7...1.8	1.8...2.0	2.0...2.3	2.0...2.5

Remark: To the lower limit of intervals given as wall height h^{**}, in the table, the value $f = 15$ cm is assigned. Intermediate values for f should be interpolated.

where n_g is the 'geometric' factor of safety with a value of 1.2; n_q is the 'load' factor of safety with a value of 1.5; and the meaning of h, q and γ was given above. In our case

$$h^{**} = 1.2 \times 2.0 + 1.5\frac{3}{20} = 2.63 \text{ m}.$$

It appears from Table 7–2 that the earth retaining wall cannot be designed with an inclination of $\Delta = 1/200$ as specified at the beginning. The requirements must be altered. In doing so, we may choose to:

— reduce the designed wall height h,
— exclude the surface load,
— moderate the inclination requirement.

Let us increase the inclination to twice its initial value; ie:

$$\Delta = 1/100.$$

Then, in the 6th column of Table 7–2, we face two possibilities: both intervals in the first and second rows encompass the value of h^{**} calculated previously. The ratio l/h (characterizing the slenderness of the earth retaining wall) can take the values 0.40 or 0.50. To demonstrate all consequences of the choice, each possibility will be investigated:

a) $l/h = 0.5$ (Fig. 7–19*a*)

The value of spacing f will be calculated by linear interpolation from the interval-limits in the table:

$$f = 30 - 15\frac{2.63 - 2.4}{3.2 - 2.4} = 25 \text{ cm}$$

(the value of f should always be rounded down). In this case the number of layers will be

$$200 : 25 = 8$$

(There is no reinforcement in the topmost layer.)

The amount of reinforcing material per meter length of the wall is:

$$2 \cdot 7 \cdot 1.0 + 7 \cdot 0.25 = 15.8 \text{ m}^2$$

b) $l/h = 0.4$ (Fig. 7–19*b*)

From linear interpolation:

$$f = 30 - 15\frac{2.63 - 2.2}{2.7 - 2.2} = 17 \text{ cm}.$$

That means that

$$200 : 17 = 12$$

reinforcing webs are needed. (The practical solution is 11·17 = 187 cm soil layer $+r$ = 13 cm overburden on top). The required reinforcing geotextile per meter length of wall in this case:

$$2 \cdot 11 \cdot 0.8 + 11 \cdot 0.17 = 19.5 \text{ m}^2.$$

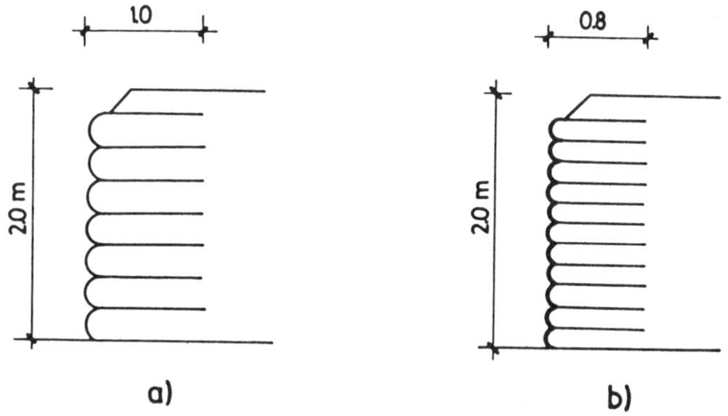

Figure 7–19 Web arrangements

To decide among these possibilities, we must take into consideration that in version *a)* for each individual layer inclusion, a length of 1.0 + 0.25 + 1.0 = 2.25 m reinforcement is needed which can be cut off from a reel of 2.0 m manufactured width only by cutting it along its length. The required overlap for splicing is 15 cm at 2.0 meters; this involves additional material use of 1.2 m^2 per meter length of face. For version *b)* the length of inclusion is 0.8 + 0.17 + 0.8 = 1.77 m. In this case the reinforcing web can be continuously rolled out in the longitudinal direction of the wall, but there will be considerable waste. If practical considerations allow the web to be rolled out in the longitudinal direction along the wall, it may be worthwhile to try to interpolate the dimension l. In this case, this would mean a third choice:

c) $$l/h = 0.45$$

The interval interpolated for h^{**} is: 2.3–2.95; then

$$f = 30 - 15\frac{2.63 - 2.3}{2.95 - 2.3} = 22 \text{ cm}.$$

The number of layers is:

$$200 : 22 = 9.$$

There will be 8 reinforced layers and 24 cm thick backfill layers.

The amount of web required per meter is

$$2 \cdot 8 \cdot 0.9 + 9 \cdot 0.22 = 16.4 \text{ m}^2.$$

As the length of each reinforcing inclusion is now $2 \cdot 0.9 + 0.22 \cong 2.0$ m, it is important to consider the saving in material due to the absence of splicing.

The construction methods for layered, geotextile or mesh reinforced earth retaining walls depends on the available plant, the features of the terrain and the transportation facilities. The general considerations are:

- the reinforcing inclusions have to be cut to size,
- fine mesh of NETLON or geotextile (depending on the sizes of the soil grains) should be laid onto the face, where grassing is also advisable (Fig. 7–20*a*),
- overlapping between adjacent meshes should be at least 15 cm,
- the topmost reinforcing layer should be extended and bent back to an angle of 30° (Fig. 7–20*b*).

General solutions cannot be given how to finish these types of retaining wall because it depends on the features of the terrain. The ends of the retaining wall depend on height and conical abutments of masonry or light concrete elements may be needed.

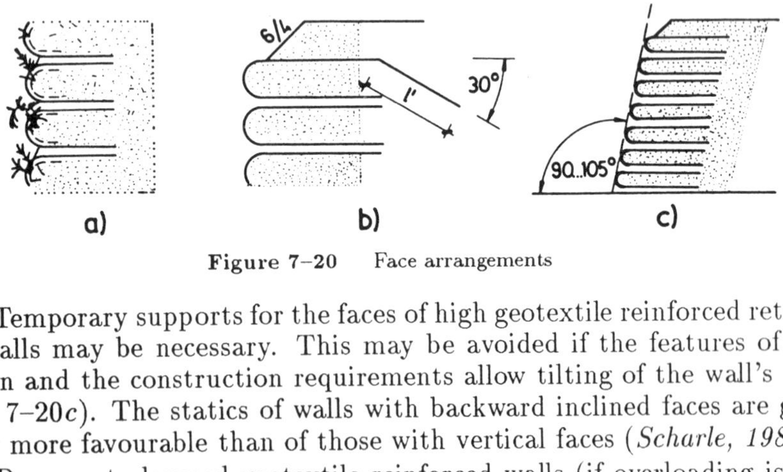

Figure 7–20 Face arrangements

Temporary supports for the faces of high geotextile reinforced retaining walls may be necessary. This may be avoided if the features of the terrain and the construction requirements allow tilting of the wall's face (Fig. 7–20*c*). The statics of walls with backward inclined faces are generally more favourable than of those with vertical faces (*Scharle, 1980*).

Damage to layered geotextile reinforced walls (if overloading is excluded) is caused principally by erosion. For this reason, when enhanced durability is sought (or perhaps always) it is worthwhile planting wall faces with vegetation. This has been discussed in Section 2.2. Fig. 7–21 shows an example of a successful construction (*Smoltczyk, Malcharek, 1981*).

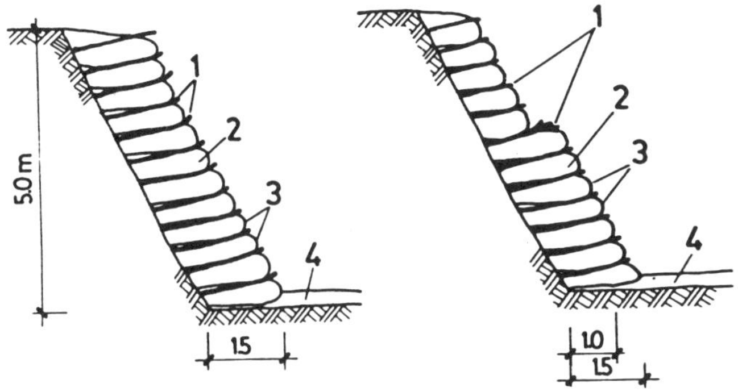

Figure 7–21 Reinforced steep slope with vegetation (*Smoltczyk, Malcharek, 1981*)
1 willow twig; *2* crushed limestone; *3* geotextile; *4* filter bed

7.3.2 Gabion walls

Box or basket shape casings filled with crushed stones, boulders, gravels or debris, are also 'earth structures' and, as gravity retaining walls, they are popular structures. Their form, construction and design are linked to the subjects described earlier and we summarize their most important features here.

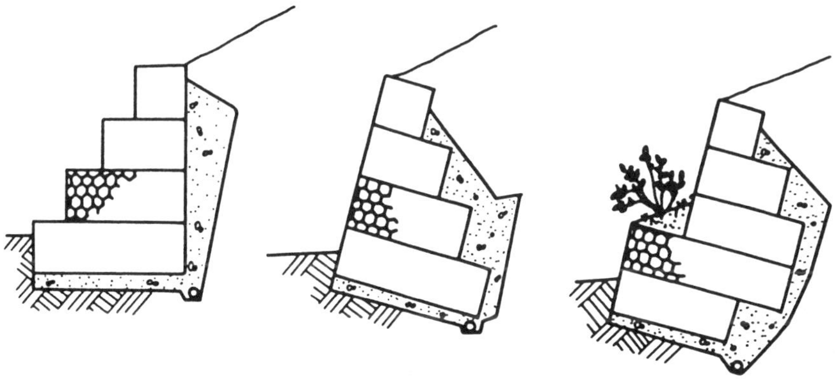

Figure 7–22 Retaining walls using gabions (*Papst, 1978*)

Some simple examples of gravity walls constructed from gabions are shown in Fig. 7–22 (*Papst, 1978*). The baskets, formed from metal- or hard plastic mesh, are filled with stones on the site. While the angle of internal friction of crushed stone would support steep natural slopes, the mesh is needed to keep the stones together to provide stable, nearly vertical, (with inclinations of 3 : 1 ... 10 : 1) wall faces.

The structure sketched in Fig. 7–23 deserves attention as a reinforced variant. The earth pressure acting on the gabion facing is decreased by the reinforced layers placed into the backfill and there is less need for costly stone material. In selecting the filling material, drainage should be kept in mind; the grain size distribution of the materials should comply with the filtering laws or the reinforcement should be supplemented by geotextiles suitable for filtering. Reinforced solutions have many advantages; their applications are spreading with the proliferation of dual purpose products (reinforcing and filtering) on the market.

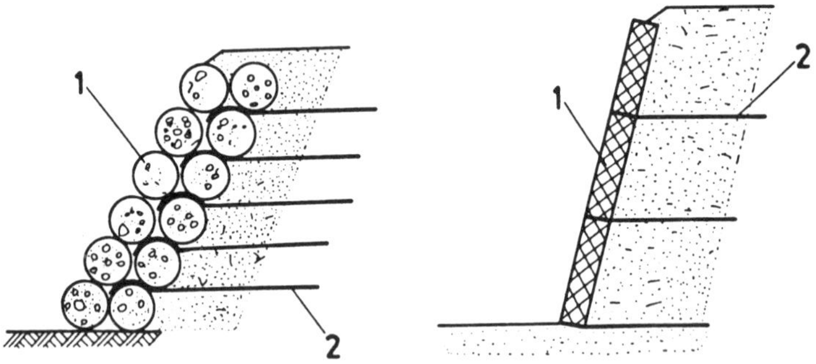

Figure 7–23 Variants which decrease the volume of gabions

1 gabion; *2* reinforcing mat or mesh

7.3.3 Light small retaining walls

For the sake of completeness, it should be mentioned that all the walls reviewed in Section 7.2 can be constructed in simple forms for use in temporary works etc. The issues are similar to those reviewed in Chapter 6. Some simple variants of these walls can be stable without a closed face covering, especially if the face is protected against weathering by plants or by a thin grouted layer.

Acceptance of these variants by clients depends mainly whether their disadvantages are offset by reduced costs. The structural simplifications are obvious and we shall not deal with them here in detail.

7.4 EARTH RETAINING WALLS BUILT OF WASTE MATERIALS

Among wastes produced by industrial processes and elsewhere, some retain their shape and strength. For example, millions of worn rubber tyres are discarded. Throughout the world, there are millions of tons of other industrial products which are difficult to recycle or dispose of.

Engineer's ingenuity can help to find rational uses for materials of this type. This section treats this question from two standpoints. On the one hand, we would like to scotch the belief that 'waste' is an absolute term and a structure of quality can only be constructed of expensive materials produced for the specific purpose. On the other hand, we would like to show that there are numerous examples of inventiveness to prove this thesis.

The materials from waste stockpiles of mines and power plants can be a good building material; and it is always worthwhile to investigate the possibility of using it for constructing earth retaining walls (*Gáspár, 1981, 1982*). In the domain of reinforcing materials, we refer to the use of plastic bands used in garden earth retaining walls (see Fig. 7–10). The reinforcing bands can be bought in rolls in lengths of several hundred meters by retaining wall contractors. For a small retaining wall in a garden, it would be better to collect and use the plastic bands of packages, which otherwise—as waste—would be discarded.

A more significant solution is presented in Fig. 7–24*a* where the cross section of a motorway embankment can be seen. The way had to be kept open at the embankments toe. It was not possible to check the stability of the steep slope where sliding and sloughing off had occurred after construction (*Dalton, 1982*). A retaining wall at the embankment's toe was unnecessary from the point of view of aesthetics. For example, a gabion wall would have been equally suitable. Earlier experiences (WC-78) induced the client to investigate the possibility of using worn rubber tyres to stabilize the slope.

The principal elements of the earth retaining wall can be seen in the axonometric view of the chosen scheme (Fig. 7–24*b*). The foundation level is 1 m below the ground surface. The position of the tyres forming the face was fixed by plastic bands and also worn tyres were used for fastening them in place. The structure is of a rather mixed character: it has the marks of a stack- (crib-) wall, an anchorage and a reinforcement. The toe of the embankment was faced by a brick wall (presumably as a protection against vandalism or with respect to aesthetics) but it has no mechanical purpose. The rims of the tyres were cut off to enlarge the circular holes to facilitate backfilling.

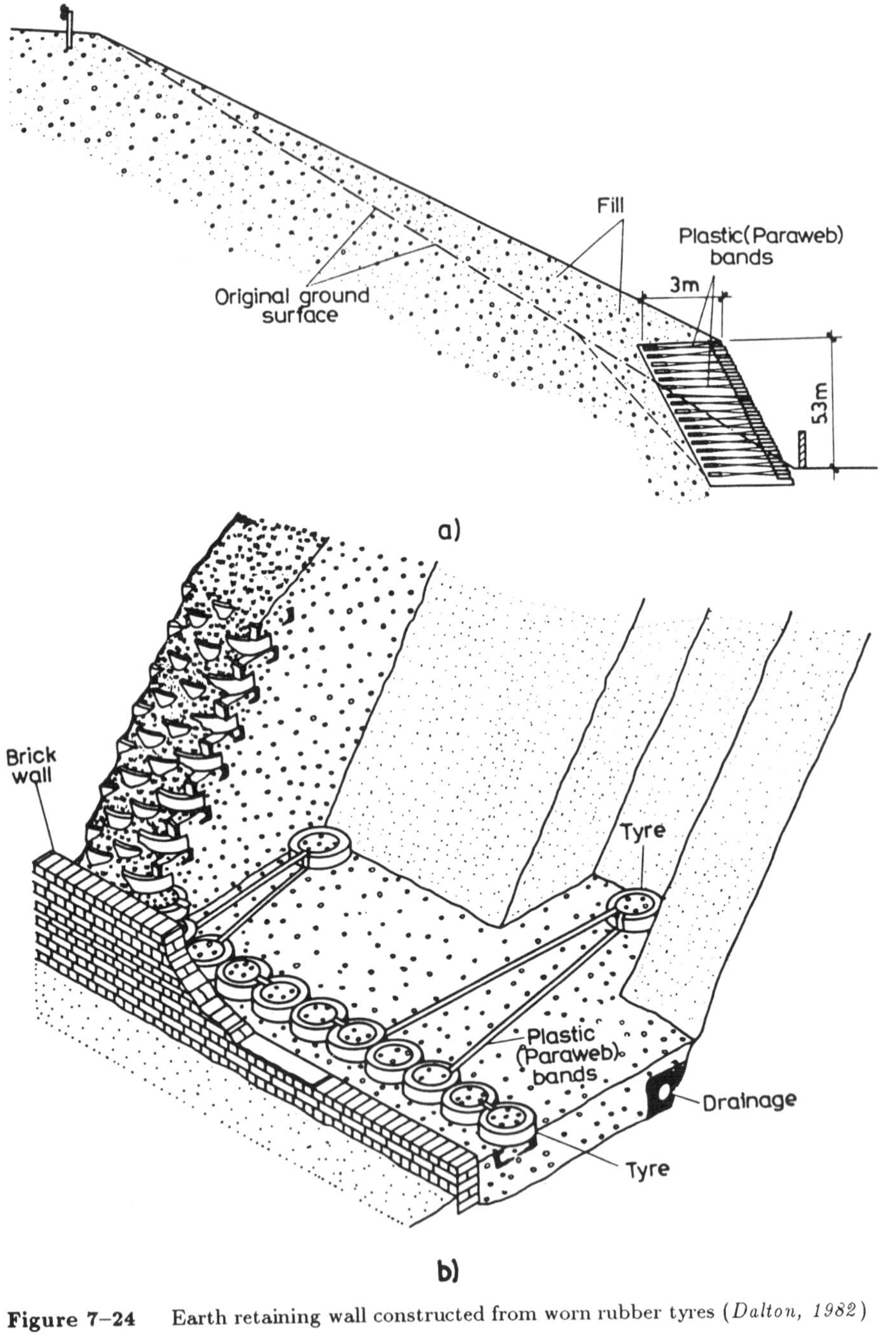

Figure 7–24 Earth retaining wall constructed from worn rubber tyres (*Dalton, 1982*)

a) section; *b)* axonometric view

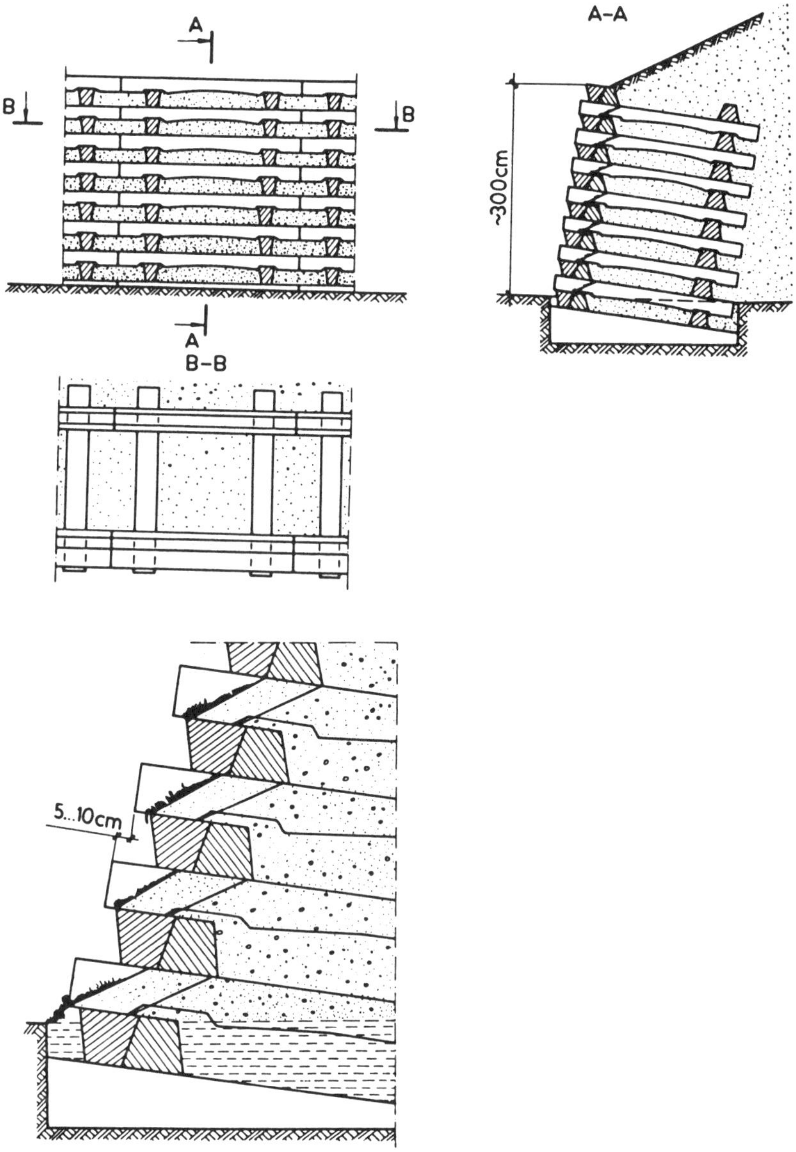

Figure 7–25 Earth retaining wall constructed from used rail tracks (*Jarominiak, 1982*)

Having constructed the wall simultaneously with the backfilling, the finished surface was grassed. The growing vegetation covered the whole surface in short time and not a single tyre can be seen.

In this case, the client obtained the tyres free of charge. About 3000 tyres were needed for a wall face of 350 m^2. Due to this saving, the client had to pay only a quarter of the costs of traditional wall structure (mainly for cutting the tyres).

The durability of an earth retaining wall constructed of rubber tyres may be 100 years. Construction method is also rapid; in this example it took only two weeks. It is worthwhile to consider whether wastes of other types (e.g. barrels) may not be used for similar structures.

Finally—to underline the last remark—another example is presented. A stack- (crib-) wall, constructed of railway sleepers (prestressed concrete beams), is shown in Fig. 7–25 (*Jarominiak, 1982*). The stock of cracked sleepers (discarded after track reconstruction) is growing from year to year. Economic use of these is of advantage to everyone.

REFERENCES

Alexander, J. M.-Brewer, R. C.-Rowe, G. W: Manufacturing Technology, E. Horwood, 1987

Alfheim, S. L.-Sorlie, A.: Testing and classification of fabrics for application in road constructions. ICUFG, 1977. 333–338.

Al-Hussaini, M. M.: Field experiment of fabric earth wall, ICUFG, 1977. 119–121.

Al-Hussaini, M. M.-Perry, E. B.: Field experiment of reinforced earth wall. ASCE GT, 1978. March. 307–322.

Andrawes, K. Z.-McGown, A.-Mashour, M. M.-Wilson-Fahmy, R. F.: Tension resistant inclusions in soils, ASCE GT, 1980. December. 1313–1325.

Bacot, J.: Methode de calcul à la rupture – Étude théorique. Service Géotechnique 999. 1975.

Balogh, A.-Galántai, M.-Kecskés, T.-Schmidt, G.: Rock-gardens, Mezőgazdasági Kiadó, Budapest, 1976 (In Hung.)

Bartos, M. J.: 101 uses for earth reinforcement, Civil Engineering. ASCE, 1979 January, 51–57.

Barvasov, V. A.-Budanov, V. G.-Fomin, A. N.-Perkov, J. N.-Pushkin, V. I.: Deformations of soil foundation reinforced with prestressed synthetic fabrics, ICUFG, 1977. 67–69.

Bányai, M.: Nailed soil retaining walls. Mélyépítéstudományi Szemle, No. 3. 97–105. (1982) (In Hung.)

Begemann, W.: Vegetative Stabilisierung von Böschungen in Lockergestein und Fels. Tiefbau, Ingenierurbau, Strassenbau. 5. 368–370. (1978)

Bell, J. R.-Steward, J. E.: Construction and observation of fabric retained soil walls. ICUFG, 1977, 123–128.

Bishop, A. W.-Henkel, D. J.: The Measurement of Soil Properties in the Triaxial Test, Edward Arnold Ltd, 1957

Bishop, A. W.: Test requirements for measuring the coefficient of earth pressure at rest, Brussels Conference on earth pressure problems, Proc. Vol. 1. p. 2., 1958

Bongartz, W.: Erste deutsche Stützwand nach dem Bauverfahren „Bewehrte Erde" bei Raunheim. Strasse und Autobahn. 5. 190–197. (1976)

Borchert, K. M.-Müller-Kirchenbauer, H.: Herrkömliche Berechnung von chemisch verfestigten Unterfangungskörpern. Die Bautechnik, 8. 275–279. (1981)

Brandl, H.: Raumgitterkonstruktionen: Tragverhalten und Neuentwicklungen. Heft 3 der Arbeitsgruppe Untergrund-Unterbau. Bonn, Kirschbaum Verlag, (without year)

Broms, B. B.: Polyester fabric as reinforcement in soil, ICUFG, 1977. 129–135.

Cartier, G.-Gigan, J. P.: Experiments and observations on soil nailing structures. ECSMF-8. 473–476.

Chang, J. C.-Forsyth, R. A.: Design and field behavior of reinforced earth wall. ASCE GT, 1977. July. 677–692.

Chang, J. C.-Forsyth, R. A.: Finite element analysis of reinforced earth wall. ASCE GT, 1977. July, 711–724.

Cookson, R. A.-El-Zafrany, A.: Finite Element Techniques for Engineering Analysis, E. Horwood, 1987

Dalton, D. C.: Tyre retaining wall on the M62. Ground Engineering, 1982 January, 41–43.

Dulcz, E.: Use of bored-injected micropiles. (Mélyépítéstudományi Szemle, No. 8. 363–367. (1981) (In Hung.)

Fejér. H.-Szalatkay, I.: Reinforced soil retaining wall implementation in Hungary. ÉGSZI Report. Városi Mélyépítés, No. 20. 26–34. (1980) (In Hung.)

Fejér, H.: Support of earth surfaces by soil nailing method. ÉGSZI Report, Városi Mélyépítés, No. 34. 10–20. (1980) (In Hung.)

Finnigan, J. A.: The creep behaviour of high tenacity yarns and fabrics used in civil engineering applications. ICUFG, 1977. 305–309.

Gáspár, L.-Czudor, L.: Utilization of garbage cinders in road construction. Mélyépítéstudományi Szemle, No. 12. 546–555. (1981) (In Hung.)

Gáspár, L.: Recent experiences with fine-ash for road construction in Hungary. Mélyépítéstudományi Szemle, No. 4. 143–150. (1982) (In Hung.)

Gässler, G.-Gudehus, G.: Soil nailing–some aspects of a new technique. ICSMF-10, 1981. 665–670.

Gässler, G.-Gudehus, G.: Soil nailing–Statical Design. ECSMF-8, 1983, 491–494.

Gedney, D. S.-McKittrick, D. P.: Reinforced earth: a new alternative for earth-retention structures. Civil Engineering, ASCE, 1975. October, 88–91.

Gerrard, C.: Reinforced soil: an orthorhombic material. ASCE GT, 1982, November, 1460–1474.

Gérusz, M.: Design and construction experiences with domestic soil nailing works. Műszaki Tervezés, No. 7. 20–22. (1981) (In Hung.)

Goldscheider, M.-Gudehus, G.: Verbesserte Standsicherheitsnachweise. Vorträge Baugrundtagung, Frankfurt/Main-Höchst, Deutsche Gesselschaft für Erd und Grundbau, 1974.

Gudehus, G.: Lower and upper bounds for stability of earth-retaining structures, ECSM-5, 1972.

Gudehus, G.: Erddruckermittlung, Smoltczyk, U.: Grundbau Taschenbuch, 3. Aufl. Berlin, Wilhelm Ernst Sohn, 1980. 327.p.

Harrison, W.J.-Gerrard, C.M.: Elastic theory to reinforced earth, ASCE SM, 1972. December, 1325–1344.

Hausmann, M. R.-Ring, G. J.: Measurement of soil/reinforcement interaction, Third Australia–New Zealand Conference on Geomechanics, 1980, Vol. 1. pp. 149–153.

Herrmann, L. R.-Al-Yassin, Z.: Numerical analysis of reinforced soil systems. ASCE Symposium on Earth Reinforcement, 1978. 428–457.

Hoare, D. J.: Laboratory study of granular soils reinforced with randomly oriented discrete fibres. ICSR, 1979. 47–52.

Holtz, R. D.-Broms, B. B.: Walls reinforced by fabrics – Results of model tests, ICUFG, 1977. 113–117.

Ingold, T. S.-Templeman, J. E.: The comparative performance of polymer net reinforcement ICSR, 1979. 65–70.

Ingold. T. S.: A laboratory simulation of reinforced clay walls. Geotechnique. 3. 399–412. (1981)

Ingold, T. S.: Soil reinforcing – current practice. Civil Engineering, April, 1982. 61–65.

Irons, B.-Ahmad, S.: Techniques of Finite Elements, E. Horwood, 1979

Jarominiak, A.: Lekkie konstrukcje oporowe, Warszava, VKL, 1982.

Jessberger, H. L.: Load-bearing behaviour of a gravel-subbase-nonwoven fabric — soft subgrace system. ICUFG, 1977. 9–13.

Jewell, R. A.-Wishatt, S. J.: Underwater construction using reinforced hydraulic fill. ECSMF-5, 1983. 847–850.

John, N. W. M.: Soil reinforcement. Civil Engineering. October, 1979. 47–53.

Jones, C. J. F. P.-Bolton, M. D.: Reinforced earth. Proc. Inst. Civ. Eng., Part I., 1977. Aug. 511–516.

Jones, C, J, F, P.: The York method of reinforced earth construction. ASCE, Symposium on Earth Reinforcement. 1978. 501–527.

Juran, I.-Schlosser, F.: Étude théorique des efforts de tractions dans les armatures des ouvrages en terre armée. ICSR, 1979. 77–82.

Юрданов, А. П.: Особенности глубинного обжига грунтов и перспективы его совершенствования, Основания, Фундаменты и Механика Грунтов. 14–16 (1978)

Юрданов, А. П.: Укрепление грунтов обжиганием, Сельское Строительство. 1. 20–21. (1980)

Karstedt, J.: Ermittlung eines aktiven Erddruckbewertes für dem räumlichen Erddruckfall bei rolligen Böden. Tiefbau, Ingenieurbau. Strassenbau, 1978, April. 258–262.

Karstedt, J.: Standsicherheitsproblem bei der Schlitzwand-bauweise. Baumaschine, Bautechnik 5, 327–334. (1980)

Kerényi, Gy.-Szalatkay, I.: Constructing of the reinforced soil retaining wall at Siófok. Mélyépítéstudományi Szemle, No. 12. 525–530. (1981) (In Hung.)

Kern, F.: Réalisation d' un barrage en terre avec pavement aval vertical au moyen de poches en textile. ICUFG, 1977. 91–94.

Kézdi, Á.: On an innovative earth retaining method. Mélyépítéstudományi Szemle, No. 12. 565–566. (1966) (In Hung.)

Kézdi, Á.: Stabilized Earth Roads. Akadémiai Kiadó, Budapest, 1967. (In Hung.)

Kézdi, Á.: Soil Mechanics. Tankönyvkiadó, Budapest, 1977. (In Hung.)

Kézdi, Á.: Practice of Soil Mechanics. Tankönyvkiadó, Budapest, 1976. (In Hung.)

Kézdi, Á.-Marczal, L.: Earth supporting with reinforced soil. Mélyépítéstudományi Szemle, No. 5. 185–188. (1979) (In Hung.)

Kézdi, Á.: Soil physics, selected topics. Budapest, Akadémiai Kiadó, 1979.

King, R. A.: A review of soil corrosiveness with particular reference to reinforced earth. TRRL, Supplementary Report, 316. (1977)

Koerner, R. M.-Welsh, J. P.: Construction and geotechnical engineering using synthetic fabrics. Chichester–New York–Brisbane Toronto, John Wiley et Sons, 1981.

Kondner, R. L.: Hyperbolic stress–strain response: cohesive soils, ASCE SM. 1963. 115–143.

Kovácsházy, F.: Design and Construction of Quays and Retaining Walls. Tankönyvkiadó, Budapest, 1966, (MTI Notes, M 130). (In Hung.)

Kovácsházy, F.–Balázsy, B.: Quays and Retaining Walls. Akadémiai Kiadó, Budapest, 1985. (In Hung.)

Kurtz, A.: Stützmauer aus Bauelementen, DT 25 15 362. 1977. (Patent)

Lee, K. L.–Adams, R. D.–Vagnernon, J. M.: Reinforced earth retaining walls. ASCE SM, 1973. October. 745–764.

van Leeuwen, J. H.: New methods of determining the stress–strain behaviour of woven and non-woven fabrics in the laboratory and in practice. ICUFG, 1977. 299–304.

Legrand, L.–Schlosser, F.: Reinforced Earth, Technical information, LCPC, Paris, Avril, 1976

Lelkes, L. (ed.) Flower and Leisure Gardens. Mezőgazdasági Kiadó, Budapest, 1982. (In Hung.)

Magyarné, J. M.–Radnóti, Gy.–Scharle, P.–Szalatkay I.: Numerical and experimental examination of the reinforced earth retaining wall in Fényeslitke, ICSR, 1979. 563–567.

Magyarné, J. M.–Scharle, P.–Szalatkay I.: Improvement of bedding conditions by inset elements. ICSR, 1979, 451–456.

Marczal, L.: Reinforced soil used for earth retaining purposes. Research report, Part II., BME Geotechnical Department, (1981). (In Hung.)

Marczal, L.: Measurements on reinforced soil constructions. Seminar on Soil Structures (Manuscript). Budapest, November, (1982). (In Hung.)

McGown, A.–Andrawes, K. Z.: The influence of non-woven fabric inclusions on the stress–strain behaviour of a soil mass, ICUFG, 1977. 161–166.

McGown, A.–Andrawes, K. Z.–Al-Hasani, M. M.: Effect of inclusion properties on the behaviour of sand. Géotechnique, 1978. 327–346.

Mitchell, J. K.: Fundamentals of soil behavior. Chichester–New York–Brisbane–Toronto, John Wiley and Sons, 1976.

Murray, R. T.: Fabric reinforced earth walls: development of design equation. Ground Engineering, Oct. 1980, 29–36.

Murray, R. T.: Oral information on about some Transport and Road Research Laboratory experiments, 1981.

Naylor, D. J.–Richards, H.: Slipping strip analysis of reinforced earth. Int. J. for Num. and Anal. Meth. in Geomechanics. Vol. 2. 1978. 343–366.

Nicholls, R.: Fabric pipelines and earth containment structures. ASCE CO. 1981, March, 9–20.

Papst, W.: Landschaftgerechtes Bauen mit Gabionen. Tiefbau, Ingenieurbau, Strassenbau, 5. 398–408. (1978).

Păunescu, M.–Donca, V.: Unele consideraţii privind eficienţ a masivelor din pămint armat aplicate la sprijinirea terasamentelor. Constructii. 5. 7–11. (1979).

Pálossy, L.: Application of industrial textiles for pipeline constructions. Építéstudományi Intézet, (1979). (Research report). (In Hung.)

Petrik, P. M.: Retaining walls anchored to reinforced soil. Mélyépítéstudományi Szemle, No. 7. 321–328. (1977). (In Hung.)

Petrik, P. M.: Projekt oporných murov z vystuženey zeminy v Margecanoch, Ročenka, 1975. Výdaná pri prilezitosti, 25. Vyrocia zalozenia Ustavu, VÚIS.

Piesker, F.–Göbel, C.: Stützbauwerke aus textilbewehrter Erde. Bauplanung-Bautechnik. February, 1982, 63–66.

Pirk, A.: Ingenious Garden Furnishing, Mezőgazdasági Kiadó, Budapest, 1977. (In Hung.)

Polgár, L.: Retaining walls constructed with precast TT panels. Magyar Építőipar, No. 4. 247–248. (1983). (In Hung.)

Postle, R.-Carbany, G. A.-deJong, S.: The Mechanics of Wool Structures, E. Horwood, 1987.

Potyondy, J. G.: Skin friction between various soils and construction materials. Géotechnique, 1961. 339–353.

Prater, E. G.: Die Gewölbewirkung der Schlitzwände, Der Bauingenieur. 4, 125–131. (1973).

Puig, J.- Blivet, J. C.-Pasquet, P.: Remblai arme avec un textile synthetique. ICUFG, 1977, 85–90.

Radnóti, Gy.-Scharle, P.-Szalatkay, I.: Plastic grid reinforced garden wall. URBENVITA '80, (1980. III/67–75.) (In Hung.)

Rainbow, A. K. M.: Use of minestone in reinforced earth construction, in Seminar of Reinforced Earth Techniques, Ground Engineering, April, 1982 (D.R.P.)

Rankilor, P. R.: Membranes in Ground Engineering. Chichester–New York–Brisbane–Toronto, Wiley and Sons, 1981.

Risse, J.-Göbel, C.-Hanspach, P.-Piesker, F.: Bewehrte Erde — eine Effektive Bauweise im Erd — und Grundbau. Bauplanung–Bautechnik, Februar, 1982. 51–53.

Romstad, K. M.-Herrmann, L. R.-Shen, C. K.: Integrated study of reinforced earth-I.: Theoretical formulation. ASCE GT, 1976. May, 456–471.

Rózsa, L.: Handbook of Foundations, 9. Soil strengthening. Műszaki Könykiadó, Budapest, 1977. (In Hung.)

Рыжсов, А. М.: Определение прочности и деформативности грунтов в строительстве, Будивелник, Киев, 1976.

Scharle, P.-Szalatkay, I.: Application and statics of reinforced earth walls. Mélyépítéstudományi Szemle, No. 7. 329–335. (1977). (In Hung.)

Scharle, P.: Dimensioning of reinforced earth constructions. Építési Kutatás Fejlesztés, No. 4. 251–256. (1980). (In Hung.)

Scharle, P.-Szalatkay, I.: Walls of reinforced earth. Building Research and Practice, 1979. July/August. 222–229.

Schiechtel, H. M.: Sicherungsbauweisen im Landschaftsbau. München, Callwey.

Schlosser, F.: La terre armée: recherches et réalisations. La techique routiére n° 3. 1972, Bulletin de Liaison n° 62. nov., dec., 1972, 79–93.

Schlosser, F.-Gobert, J.: La terre armée dans l'échangeur de Séte. Revue Générale des Routes et des Aerodromes, oct. 1972. Bulletin de Liaison n° 63. jan., fév., 1973. 149–171.

Schlosser, F.-Juran, I.: Design parameters for artificially improved soils, DPGE, 1981, V. 5. 197–225.

Schlüter, K.: Die Eignung von Holzarten für den Busch- und Heckenlagenbau. Landschaft + Stadt. Beiheft, 6. (1971).

Schwab, E. F.-Pregl, O.-Broms, B. B.: Deformation behaviour of reinforced sand at model tests measured by X-Ray. ICUFG, 1977. 105–112.

Shen, C. K.-Romstad, K. M.-Herrmann, L. R.: Integrated study of reinforced earth-II.: Behavior and design. ASCE GT, 1976, June. 577–590.

Siedek, K.: Böschungssicherung. Strassen- und Tiefbau, 19. (1965).

Sissons, C. R.: Strength testing of fabrics for use in civil engineering. ICSR, 1979. 287–292.

Smith, A. K. C.-Bransby, P. L.: The failure of reinforced earth walls by overturning. Géotechnique. 2. 376–381. (1976).

Smith, G. N.-Birgisson, G. J.: Increasing the strength of reinforced earth retaining walls by inclined strips. Civil Engineering, June. 1979. 62–63.

Smoltczyk, U.-Malcharek, K.: Lebendverbau von Steilböschungen. Tiefbau, Ingenieurbau, Strassenbau, 1981. 398–400.

Snaith, M. S.-Bell, A. L.-Dubois, D. D.: Embankment construction from marginal material. ICSR, 1979. 175–180.

Sohnius, D.: Zugkraftmessungen mit Dehnungsmessstreifen an Bewehrungsbändern einer Stützkonsruktion nach dem Bauverfahren ,,Bewehrte Erde". Messtechnische Briefe 18. 1982. Heft 2, 32–36.

Sorlie, A.: The effect of fabrics on pavement strength. ICSR, 1979. 15–18.

Stocker, M.-Gässler, G.: Ergebnisse von Grossversuchen über eine neuartige Baugrubenwand-Vernagelung. Tiefbau, Ingenieurbau, Strassenbau, 9. 677–682. (1979).

Szabó, I.: Steel Fibre Reinforced Concrete. Müszaki Kiadó, Budapest, 1976. (In Hung.)

Szalatkay, I.: Reinforced soil retaining wall. Tender plan. Építéstudományi Intézet, Budapest, (1975). (Documentation, 804/75.) (In Hung.)

Szalatkay, I.: Reinforced soil retaining walls and their static analysis. MSc theses. Budapest, (1977). (In Hung.)

Szalatkay, I.: Elastic properties of the reinforced earth. Acta Technica. 86. 215–229. (1978).

Szalatkay, I.: Sopron, Templom u. 12. Construction plans of a reinforced soil retaining wall. Építéstudományi Intézet, Budapest. (Documentation, 1319/78.) (In Hung.)

Szalatkay, I.: Replacement of steel strips by plastic grids in reinforced soil retaining walls. Építéstudományi Intézet, Budapest, (1981). (Documentation.) (In Hung.)

Szalatkay, I.: Reinforced soil wall on the road No. 61. Építéstudományi Intézet, Budapest, (1983). (Report.) (In Hung.)

Szalatkay, I.: Design and construction of reinforced soil retaining walls. TS–M–57. TTI, (1983). (In Hung.)

Szalatkay, I.-Vas, J.: Advanced applications of Netlon grid reinforced soils. Dimensioning of soil walls using experimental results. Építéstudományi Intézet, Budapest, (1983). (Documentation.) (In Hung.)

Széchy, K.: Foundations II. Müszaki Könykiadó, Budapest, 1963. (In Hung.)

Toth, L.: Gewölbewirkung bei einer aufgelösten Bohrpfahlwand und Ermittung der lichten Abstände der Bohrpfähle. Die Bautechnik. 6 181–188. (1978).

Vidal, H.: La terre armée. AITBTP, 1966. 888–938.

Vidal, H.: La terre armée. AITBTP, Supplement, 1969. 1099–1156.

Vidal, H.: Reinforced Earth. AITBTP, 1972. 139–166.

Walkinshaw, J. L.: Reinforced Earth Construction, Final Report, Dep. of Transp. FHWA–DP–18, April, 1975.

Walz, B.-Prager, J.: Beitrag zur Lösung von Erddruckproblemen nach der Elementscheibentheorie. Die Bautechnik. 11, 12, 373–377., 424–427. (1979).

Warwick, E. R. J.: Reinforced earth — its design and application. Civil Engineering, Sept. 1982. 51–59.

Woodtli, K.: Erfahrungen mit Lebendverbaumethoden in der Schweiz. Strasse und Verkehr. 4. 134–139. (1978).

Yamanouchi, T.: Resinous net applications in earth works, Conference on Soil Stabilization and Compaction. Sydney, 1975. 5.1.–5.16.

Yamanouchi, T.-Miura, N.-Matsubayashi, N.-Fukuda, N.: Soil improvement with quicklime and filter fabric. ASCE GT, 1982 July. 953–965.

CDM-RSS: A computer design method for reinforced soil structures using limit equilibrium analysis. TN/8292/A. London, Binnie and Partners, 1981.

Begrünter Schallschutz aus Beton-Profilen (*dabau*). Strassen- und Tiefbau, 1976/11, 34. (Information)

Fodervävnader AB: A system for reinforcing of earth, PS 1 453 573, 1976. (Patent)

LOTAR: *Les ouvrages en terre armée.* Recommendations et régles d'art. Ministere des Transports, September, 1979.

NEW: *Neue Ebenseer Wand,* Sonderbau GmbH, Wien (product manual, without year).

RECC: *Reinforced earth cuts costs.* Civil Engineering, May 1976. 9. (Advertisement)

RETT: *Reinforced earth tried and tested* for over ten years. Civil Engineering, March 1976. 9. (Advertisement)

RE-TM: *Reinforced earth* retaining walls and bridge abutments for embankments. Department of Transport, Technical Memorandum, BE 3/78.

TAM: *Terre armée dans le monde.* 1977. ("Terre armée" leaflet, without year)

VR-ABBE: *Vorlaufige Richtlinien* für die Anwendung des Bauverfahrens "Bewehrte Erde". 1977. Betonkalender, 1978/II.

WC-78: *World Construction,* March, 1978. (Information matter)

WC-80: *Fabrics for overlaying* pavements, low-cost retaining walls, solving drainage problems. World Construction. June, 1980.

Abbreviations in the References

AITBTP: Annales de l'Institut Technique du Batiment et des Travaux Publique.

ASCE: Proceedings of the American Society of Civil Engineers, Journal of the
CO—Construction.
GT—Geotechnical Engineering.
SM—Soil Mechanics and Foundation.

ECSMF-5: Proceedings of the 5th European Conference on Soil Mechanics and Foundation Engineering, Madrid, 1972.

ECSMF-7: Proceedings of the 7th European Conference on Soil Mechanics and Foundation Engineering, Brighton, 1979.

ECSMF-8: Proceedings of the 8th European Conference on Soil Mechanics and Foundation Engineering, Helsinki, 1983.

ICSMF-10: Proceedings of the 10th International Conference on Soil Mechanics and Foundation Engineering, Stockholm, 1981.

ICUFG: International conference on the use of fabrics in geotechnics, Paris, 1977.

ICSR: International conference on soil reinforcement. Paris, 1979.

SICG: Second international conference on geotextiles. Las Vegas, 1982.

SUBJECT INDEX